Soil and Water Science: Key to Understanding Our Global Environment

Related Society Publications

Agricultural Ecosystem Effects on Trace Gases and Global Climate Change

Defining Soil Quality for a Sustainable Environment

Impact of Carbon Dioxide, Trace Gases and Climate Change on Global Agriculture

For information on these titles, please contact the ASA, CSSA, SSSA Headquarters Office; Attn: Marketing; 677 South Segoe Road; Madison, WI 53711-1086. Phone: (608) 273-8080. Fax: (608) 273-2021.

Soil and Water Science: Key to Understanding Our Global Environment

Proceedings of a symposium sponsored by Divisions S-1 and S-6 of the Soil Science Society of America. The symposium was held in Cincinnati, OH, 10 Nov. 1993. The symposium was held in honor of Daniel Hillel.

Convenor
Ralph S. Baker

Editorial Committee
Ralph S. Baker
Glendon W. Gee
Cynthia Rosenzweig

Managing Editor
David M. Kral

Associate Editor
Marian K. Viney

SSSA Special Publication Number **41**

Soil Science Society of America, Inc.

Madison, Wisconsin, USA

1994

Cover Design: Patricia J. Scullion
Front cover color photo taken from Chapter 7 by Cynthia Rosenzweig. Planet Earth viewed from the Apollo Mission (Courtesy of NASA, Washington, DC).

Soil Science Society of America, Inc.
677 South Segoe Road, Madison, WI 53711 USA

Library of Congress Cataloging-in-Publication Data

Soil and water science : key to understanding our global environment / editorial committee, Ralph S. Baker, Glendon W. Gee, Cynthia Rosenzweig.
p. cm. — (SSSA special publication : no. 41)
"Proceedings of a symposium sponsored by Divisions S-1 and S-6 of the Soil Science Society of America . . . held in Cincinnati, OH, 10 Nov. 1993 . . . in honor of Daniel Hillel."
Includes bibliographical references.
ISBN 0-89118-816-9
1. Soil science—Contresses. 2. Arid regions agriculture--Congresses. 3. Water-supply—Congresses. 4. Soil and civilization—Congresses. 5. Water and civilization—Congresses. 6. Irrigation—Contresses. 7. Agriculture—Congresses. I. Baker, Ralph S. II. Gee, Glendon W. III. Rosenzweig, Cynthia. IV. Hillel, Daniel. V. Soil Science Society of America. Division S-1. VI. Soil Science Society of America. Division S-6. VII. Series.
S590.2.S624 199
333.73—dc20 94-31646
CIP

Printed in the United States of America

CONTENTS

FOREWORD

Soil and water have played a critical role in the development and advancement of civilization. Where sound management of these resources was practiced, civilization flourished; where short-sighted or wasteful policies were employed, it declined. All too often the lessons of history have been forgotten or ignored by the present occupants of our planet, either because the message was too subtle, or because the wisdom of the teacher was not recognized. In combining the study of historical patterns of natural resource use with the application of the principles of soil and hydrologic sciences, Daniel Hillel has made history's message absolutely clear: Soil and water are precious and fragile resources that are vulnerable to mismanagement, and short-sighted policies will have far-reaching and disastrous consequences. This publication arises out of a special symposium, *Soil and Water Science: Key to Understanding our Global Environment*, held at the annual meeting of the Soil Science Society of America in 1993. Its unique blending of history and contemporary issues in soil and water management has created a valuable and fascinating document that should be required reading for all soil scientists. For as Santayana stated so well: "Those who forget the past are condemned to repeat it."

WILLIAM A. JURY
Soil Physics Division S-1 Chair 1993

Daniel Hillel

PREFACE

A festschrift, as defined in the dictionary, is a volume of learned articles or essays by colleagues and admirers, serving as a tribute to an esteemed scholar. This volume is, fittingly, a festschrift dedicated to Professor Daniel Hillel, a scholar, researcher, educator, author, and adviser to international development agencies. The breadth and quality of Dr. Hillel's contributions span four decades, six continents, and a dozen disciplines. His work has ranged from arid zone ecology to tropical soil management; from theoretical studies of soil-water-plant dynamics to practical innovations in the control of infiltration, evaporation, tillage, soil structure, and irrigation. His extensive publications (more than 200 papers and research reports) are noted for their clarity and originality. Especially distinguished are his 15 books, through which he has reached and educated thousands of students throughout the world. Dr. Hillel has, in fact, represented our science with such eloquence as to become not merely the spokesman for soil physics, but indeed its poet laureate.

This volume, however, does not pretend to honor Daniel Hillel's entire career. Since he is fully active, and intends to continue to be so, the time is not ripe for us to encapsulate his work *in toto*. Rather, we focus here on one of Hillel's recent contributions, his landmark book *Out of the Earth: Civilization and the Life of the Soil*, published in 1991 by The Free Press (Macmillan), and later republished by Aurum Press and the University of California Press.

More specifically, the volume herewith comprises the proceedings of a symposium on "Soil and Water Science: Key to Understanding our Global Environment." That symposium was convened in honor of Daniel Hillel as part of the Annual Meetings of the American Society of Agronomy-Soil Science Society of America-Crop Science Society of America. It was held in Cincinnati, OH, on 10 Nov. 1993, and was attended by an audience of more than 300. Coming as it did about six months following Dr. Hillel's retirement from the University of Massachusetts, where he had served for some 17 years as professor of soil physics and hydrology, the symposium was an occasion in which many authors and members of the audience took great pleasure in honoring their friend, colleague, and teacher.

The symposium was informed and inspired by the book *Out of the Earth: Civilization and the Life of the Soil*, and by its inquiry into the history and the contemporary state of our relationship with the earth. In this work, written in his distinctive style, Hillel has interwoven science with history, folklore, philosophy, and even theology, to explore the relevance of our science to the larger issues of society and the global environment. He has thereby reached beyond the confines of our profession to the wider public, thus promoting broader appreciation of soil and water as the fundamental resources upon the wise use of which hangs the fate of our civilization. The book was selected by the American Association of Publishers as the outstanding work of scholarship (first place award) in geography and the earth sciences for 1992.

Appropriately, the symposium focused on Dr. Hillel's concern over the past and present uses and misuses of the environment in the light of soil and hydrological science. It was the organizer's hope that the perspectives elucidated might shed further light on the reasons for past failures, and offer solutions to current problems in land and water management.

Daniel Hillel's keynote address set the tone for the papers to follow by underscoring the growing recognition that environmental science must transcend the limitations of the classical reductionist approach to inquiry and forge links with sister physical and biological sciences, as well as with the heretofore separated concepts of the social sciences and humanities. Hillel thus calls for agronomists and soil scientists to build bridges to disciplines, institutions, and societies outside our habitual spheres of activity so that together we may find the wisdom to meet today's urgent environmental challenges. Likewise, he asks that we adopt a more inclusive attitude, recognizing ourselves as but one among many species sharing and constituting the community of life on earth, with neither the right nor the responsibility to manipulate and endanger the natural biosphere.

Robert McCormick Adams' chapter traces the fateful interplay between soil and water resources in the Fertile Crescent and the adaptation or maladaptation of early societies to those resources. Historical evidence reveals marked differences among patterns of land use in the Nile and the Tigris-Euphrates valleys, differences attributable to the variability and reliability of the flows in each of the rivers. Ancient irrigators had the same tendency as modern irrigators to overwater when supplies were abundant, and thus to induce waterlogging and salinization in the absence of adequate drainage.

Montague Yudelman, an economist, shows that while irrigation has been among the leading contributors to the increase in food production in the developing countries, prospects for continued expansion of irrigation will be increasingly limited. In many parts of the world, the most favorable sites for irrigation projects have already been used. Improved management of existing systems, rather than primary reliance on developing ever new systems, is called for.

John Letey's perspective on the question "Is irrigated agriculture sustainable?" incorporates physical and biological, as well as societal and economic, factors. Although the science of irrigation is well established and knowledge of how to minimize salinization is available, short-term societal and economic considerations perpetuate practices that lead to eventual degradation. Once the longer-term costs of inefficient water use, loss of soil resources, and diminishing productivity are factored in, however, investments in more efficient irrigation methods become feasible and indeed inescapable.

The global overview of soil erosion presented by Rattan Lal underscores the enormity of the problem. Sadly, however, the accuracy and reliability of the relevant data leave much to be desired. Lal shows that while some quantitative data are available, they often cannot be compared across regions, nor from one scale to another. These inadequacies must be overcome in order that urgent management decisions can be made on the basis of sound knowledge.

Harold Dregne's focus on land degradation in arid regions incorporates soil erosion, salinization of irrigated land, and the effects of overgrazing and woodcutting on the loss of vegetative cover. Although most of the world's arid rangeland suffers from overgrazing, much of this damage may well be reversible. Estimates of the loss of economic productivity associated with the degradation of arid lands are constrained by the paucity of experimental measurements.

Cynthia Rosenzweig reviews the evidence regarding global warming and its possible impacts on agriculture. In a warmer world, many expensive adjustments will be necessary. Presently cool areas may well benefit, while in warm areas rising temperatures may exceed the tolerance of some crops and potential productivity may be further reduced by the hastened maturation of crops. Although the increased concentration of carbon dioxide can promote photosynthesis and water-use efficiency, some inland areas will probably suffer greater moisture deficits resulting from higher evaporative demand. Low-lying areas along coasts may be affected by sea-level rise and salt-water intrusion. Climate changes will also affect the distributions of diseases, weeds, and insect pests.

Herman Bouwer makes a compelling case that future urban water management strategies will need to rely increasingly on the re-use of treated wastewater following its discharge into designated soil-aquifer recharge basins. Such recycling affords both storage and passive treatment (geopurification) of wastewater sufficient for many nonpotable uses, while allowing the preservation of high quality aquifers for potable uses. With increasing demands being placed on finite water resources, reconciliation of conflicting interests will require the adoption of cooperative, innovative, and cost-effective approaches to wastewater disposal and reuse.

In Jan van Schilfgaarde's perspective, American society's growing awareness that agriculture and water development have resulted in numerous environmental problems has been accompanied by significant advances in fundamental knowledge, achieved by soil physicists and other scientists. Yet this knowledge has not by and large been translated into socially acceptable, practical solutions. We must therefore rise to the challenge of applying our knowledge to find more sustainable ways to manage our natural resources.

Nyle Brady examines the achievements and challenges of developing countries and of international programs in the sphere of agriculture and natural resource management. While the green revolution made great strides in alleviating hunger in many countries, future agricultural development must be carried out with greater attention to environmental sustainability. New science-based technologies, employing advances in genetics, agronomy, engineering, plant nutrition, and multifactor modeling will be necessary. Commensurate attention must be paid to social and cultural compatibility, to ensure that third world farmers collaborate in the initiatives that scientists and technologists offer them.

Within these chapters are found some of the principles that our honoree advocates: recognition of the multidisciplinary nature of environmental and agricultural problems, awareness that sustainable stewardship must replace careless exploitation, and transition from exclusive self-concern toward ever-expanding radii of inclusion. We are grateful to Dr. Hillel for elucidating these concepts and inspiring us to follow his example of accomplishment as we go forward in our individual careers in soil science and in our collective endeavor as stewards of the earth and its resources.

Ralph S. Baker

Glendon W. Gee

Cynthia Rosenzweig

CONTRIBUTORS

Robert McC. Adams	Secretary, Smithsonian Institution, Washington, DC 20560
Ralph S. Baker	Technical Director, ENSR Consulting and Engineering, Acton, MA 01720
Herman Bouwer	Chief Engineer, USDA-ARS, U.S. Water Conservation Laboratory, Phoenix, AZ 85040-8832
Nyle C. Brady	Senior International Development Consultant, United Nations Development Programme and the World Bank, Washington, DC 20006
H. E. Dregne	Horn Professor Emeritus, International Center for Arid and Semiarid Land Studies, Texas Tech University, Lubbock, TX 79409
Glendon W. Gee	Hydrology Section, Battelle Pacific Northwest Laboratories, Richland, WA 99352
Daniel Hillel	Professor Emeritus, Department of Plant and Soil Sciences, University of Massachusetts, Amherst, MA 01003
Rattan Lal	Professor of Soil Science, School of Natural Resources, The Ohio State University, Columbus, OH 43210
J. Letey	Professor of Soil Physics, Department of Soil and Environmental Sciences, University of California, Riverside, CA 92521
Cynthia Rosenzweig	Associate Research Scientist, Columbia University and NASA Goddard Institute for Space Studies, New York, NY 10025
Jan van Schilfgaarde	Associate Deputy Administrator, USDA-ARS, BARC-West, Beltsville, MD 20705
Montague Yudelman	Senior Fellow, World Wildlife Fund, Washington, DC 20037-1175

1 Introductory Overview: Soil, Water, and Civilization

Daniel Hillel

University of Massachusetts
Amherst, Massachusetts

In my youth, I was an inquisitive adventurer. Four decades ago, after having gained a master's degree in soil science and regarding myself a "compleat" scientist, destined for great discoveries, I set out to explore the legendary deserts of the Middle East. I devoted several years to studying the natural habitats and the possibilities for agricultural development in the Negev and Sinai. As part of my study, I lived for a time with a tribe of nomadic Bedouins, who then lead an extremely austere life grazing emaciated goats and camels on the sparse vegetation of the desert range. I spent many days with the herders and their flocks, wandering over winding wadis and rocky hillsides.

One day I sat with the tribe's children at the feet of their venerable old *mualem* (teacher) as he was instructing his charges in *turuk el adab*, the proper *ways of the world.* The man had no formal schooling, only the wisdom of the ages to impart to the young generation. And he did so in the time-honored way of the sages, by parable and analogy, question and answer. "How much is one and one?" asked the venerable instructor. "Two," ventured the eldest of the children. "Sometimes it is so, but sometimes not," replied the *mualem*. Then he explained: "If you put one she-goat in a pen and then another, surely you will have two. But if you put one he-goat with another, they will fight and one might kill the other so that one and one would end up being one; or they might kill each other so that one and one would be none. However, if you put together one she-goat and one he-goat, you might eventually have three, or four, or more!" So, concluded the worldly-wise old teacher: "How much is one and one? That depends on circumstances and on inclinations. Nothing is certain. And so it is with human beings, whether good or bad. And may Merciful Allah provide for the good..."

At first I was amused by that lesson, but thought nothing more of it. As I became a research scientist, I disdained uncertainty. I disciplined myself to value hard facts and to seek exact quantitative relationships among clearly definable factors. Like so many of my colleagues, I designed experiments and tested theories based on the classical premise of *ceteris paribus*, all else being equal. If only we could define single processes in a "clean" environment (such as a system of glass beads in a constant-temperature laboratory) unencumbered by extraneous complications, we should be able to formulate equations for the functional

---

Soil and Water Science: Key to Understanding Our Global Environment, SSSA Special Publication 41.

relations between dependent and independent variables. Eventually, having defined *individual* processes separately, we should also be able to add them together so as to characterize the entire natural system.

Only gradually did I learn that, in the real world, *ceteris non paribus* is the norm. The old paradigm of reductionist science did not apply very well to complex natural systems, particularly not to living biological systems.

With experience and maturity come realizations. In the course of time I recalled and belatedly understood the wisdom of the old Bedouin master. His dictum encapsulated an ecological truth with profound implications. In Euclid's simplistic axiom, the whole is equal to the sum of its parts: one plus one always equals two. That is so, however, only when the parts are inert, noninteractive, sterile. Not so in an ecosystem, where the whole includes not merely the sum of discrete parts but, also the gamut of their complex interactions, both synergistic and antagonistic. In a living community of plants and animals (including humans), each entity is defined not alone by its individual characteristics, but also by its relationships and interactions with all other entities sharing the same domain.

The science of the 19th and early 20th centuries sought to separate the world into precisely definable entities. In the process, it also classified and separated scientists into distinct categories: chemists, physicists, and biologists. Those categories, in turn, were subdivided into subcategories, and so *ad infinitesimum*. Consequently scientific institutions became compartmentalized, as individual scientists burrowed into their exclusive niches with particular jargons and separate publication media. That divergence led to its diametric opposite: a holistic attempt to characterize the entire system without sufficient knowledge of its component parts and processes. An apt joke was that science had been split between specialists and generalists: the specialists would study more and more about less and less, until, ultimately, they would know everything about nothing; while generalists would learn less and less about more and more until they would know nothing about everything.

Increasingly, environmental science requires that we bridge that artificial gap by studying more and more about more and more. The science of the 21st century is likely to continue this trend. It will explore increasingly complex relationships and interactions. And since single human beings are typically limited in perception and comprehension, the new paradigm of environmental research will require team work. So, we can safely predict, scientists will need to forgo their prideful status as autonomous *primo dons* and *prima donnas* in order to form complementary teams encompassing the physical and biological sciences. Surely, that will not be easy. In the process, we may need to sacrifice the certitude of working in isolated, narrowly defined systems in order to gain relevance to natural systems, which are by the nature of their complexity less easily predictable and hence, some might say, messier.

Even more difficult will be the task of building bridges across the wider and deeper trench dividing the natural sciences from the so-called social sciences, the dichotomy that C.P. Snow called "the two cultures." This too, however, is an inescapable challenge. All phenomena take place in context, and so does scientific research. Though it requires the best efforts of individuals, it is indeed a collective social undertaking. And to the extent that it is directed toward the resolution of actual problems (such as anthropogenic ecological degradation and global warming), it must be based on an awareness of the historical and societal roots of these problems. Knowledge of the past is essential to an understanding of the present, and to an intelligent forecasting of the future. It is the nugget of wisdom that may lie hidden in the ashes of failed prior civilizations.

If we as environmental scientists focus our attention exclusively upon the predicaments of the moment, we are likely to find ourselves repeatedly surprised and buffeted by a host of bewildering problems seeming to come out of nowhere, without a past and hence without direction. How did these problems arise? Chances are, the seeds of the phenomena we witness today were planted some time ago by our predecessors, as indeed we are planting the seeds of future problems, perhaps unknowingly, at this very moment.

What, however, is history? Though there is only one past, which cannot be changed, it can be variously presented and interpreted. Skeptics say that history is what historians say happened at particular places and times. And who are those historians? By and large, they are fallible humans who set themselves to the task of sifting and winnowing through the records of the past, which are notoriously incomplete and nebulous, in the fervent hope of discovering some explanation for how and why events happened as they apparently did.

According to one particularly cynical view, expressed by the American sociologist and economic historian William Graham Sumner (1840–1910): "All history is one long story to this effect: Men have struggled for power over their fellow men in order that they might win the joys of earth at the expense of others, and might shift the burdens of life from their own shoulders upon those of others."

Constant struggle is indeed the hallmark of history, yet historians differ on the crux of that struggle. Was it power for power's sake or for territorial or economic gain? And why did some groups win over others? Was their victory due to sheer luck, to better tactics and organization, or to superior technology? Or were the winners those who were driven by greater conviction or religious fervor? All these factors and more must have played a role in the turbulent course of history.

Classical historians were taught history as a discipline in itself. They were rarely trained in science and hardly ever in environmental science. Hence they tended to focus upon clashes among tribes, nations, dynasties, religions, or economic classes; variously led by kings, generals, or charismatic leaders. They argued among themselves whether to ascribe the decline of one society or another (or the defeat of one society by another) to such factors as moral decay, cultural enfeeblement, lead poisoning, lack of military preparedness, or what have you, when in fact the main contest for a society's survival or ascendancy had really been decided internally by the deterioration of its environmental base.

Absent until recently from the study of history was the crucial role of environmental factors in the rise and demise of societies. It is only in the last few decades that the ecological dimension has been recognized explicitly by a few pioneering historians and archaeologists, notably including such people as C.A. Wittfogel, Karl W. Butzer, Donald Worster, Thorkild Jacobsen, and in particular, Robert McCormick Adams (author of Chapter 2). They began to appreciate that the fate of ancient societies rested upon their management or mismanagement of resources, especially the most basic resources that are particularly relevant to us here: soil and water. One might even say that, in a profound sense, the essence of history is the manner by which groups of humans adapted their lives to their environment, and what is increasingly important in the course of civilization, how they modified their environment to suit their own purposes.

Before I began my study of human interactions with the environment, I had held the rather prevalent idea that human abuse of the environment is a new phenomenon, mostly a consequence of the recent population explosion and of our modern technological and materialistic economy. Ancient societies, I presumed, were more prudent than ours in the way they treated their resources. For the most part, that has turned out to be a romantic fiction. My research has led me to the conclusion that manipulation and modification of the environment was a characteristic of many societies from their very inception. Long before the advent of earth-moving machines and toxic chemicals, even before the birth of agriculture, humans began to affect their environment in far-reaching ways that destabilized natural ecosystems.

The Middle East is a particularly interesting region in which to study the environmental history of civilization. Here, in the wake of the last ice age some 10 000 yr ago, our predecessors made the fateful transition from their earlier nomadic existence as hunter-gatherers to become permanent herders and cultivators. Here they first domesticated plants and animals, tilled and irrigated fields and orchards, established villages and cities, built temples and monuments, and organized nation states that grew into empires. Here they invented ceramics, metallurgy, mathematics, and writing. And here they conceived and enunciated universal ethical and religious ideals, and codified them into laws. Indeed, the record of their activity is etched into the face of the land.

Seeing the ravaged state of much of the Middle East today, one wonders how all that could have taken place in this seemingly inhospitable corner of the earth. The fabulous oil-based wealth of the few only accentuates the deprivation of the many. The contrast between the brilliant record of the past and the depressed condition of the present demands an explanation.

The explanation is environmental. The Middle East is where we find the most dramatic and tragic examples of human-induced environmental degradation. The poor condition of this region today is due in large part to the prolonged exploitation of its fragile environment by generation after generation of cutters and burners of forests, grazers and overgrazers of rangelands, cultivators of slopes, and irrigators of river valleys, all diligent and well intentioned, but cumulatively destructive nonetheless. The wounded environment is the real reason why this region is now seething with unrest and violence. Throughout the region, the traditional village-based society is being undermined by the degradation of the soil and water, coupled with the uncontrolled population explosion. Millions of people, displaced from the land, aggregate in cities without infrastructure or productive employment, hence they are understandably resentful of modernity and are prone to the instigation of violence, of which they are the primary victims. Their fate, in fact, is duplicated in parts of Africa as well as of the Americas.

As the cradle of Western Civilization, the Middle East actually witnessed the sequential development of three forms of society, or civilization, which coexisted in competitive interdependency. The first of these was based on rain-fed farming in the relatively humid arc of highlands and intermontane valleys that gird the Fertile Crescent on the west, north, and east. It began with the domestication of local plants, such as annual cereal grains and legumes, and later fruit trees. The second civilization extended into the semiarid savannah-like interior rangelands, and was based on the herding of domesticated animals such as sheep and goats. The third and most intensive agricultural civilization was somewhat later in coming, it extended farming from the rainfed highlands to the arid valleys, and was based on the irrigation of alluvial soils by the natural flooding or artificial diversion of water from rivers.

Each of these civilizations degraded the land in its own way. Rainfed farming eradicated the tree and shrub cover and pulverized the soil, thus inducing accelerated erosion. Pastoral seminomadism tended to overgraze the sparse vegetation of grasses and bushes, thus baring the dry soil to wind erosion and increasing its vulnerability to water erosion under the infrequent but intense rainstorms. Together, these processes of land degradation (today called *desertification*) created desert-like conditions in semiarid areas. Finally, irrigated farming in ill-drained river valleys caused waterlogging and salinization.

Although the Middle East is a particularly poignant example of long-term environmental degradation, it is by no means the only example. The cultivation and grazing of sloping terrain without effective soil and water conservation took place throughout the Mediterranean region (including parts of southern Europe and north Africa), which has borne the brunt of human activity for some five millennia. Consequently, the original mantle of fertile soil, perhaps 1 m deep, that covered the hills was raked off by the rains and deposited in valleys, where the sediment clogged estuaries and impeded drainage. Widespread land degradation may well have been the reason why the Phoenicians, Carthagenians, Greeks, and Romans, in turn, were compelled to venture away from their own countries and to establish far-flung colonies in pursuit of new productive land. The end came for each of these empires when it had become so dependent on faraway and unstable sources of supply that it could no longer maintain central control.

There were, on the other hand, a few societies that did better than others. Some ingenious and diligent societies developed modes of life and methods of operation that enabled them to thrive in difficult circumstances for many centuries. Judicious management of soil and water is exemplified in some of the arid regions of the Middle East and the American Southwest. Equally impressive is the evidence regarding the long-lasting wetlands-based societies of Meso-America and South America. Remarkably productive wetland management systems have survived intact in China and other parts of Southeast Asia. In contrast with the historic failures of Mesopotamia and the Indus Valley, the irrigation-based civilization of Egypt sustained itself for more than five millennia, though it is now beset with problems of unprecedented severity.

Every one of the insidious man-induced scourges that played so crucial a role in the decline of past civilizations has its mirror image in our contemporary world. But it seems that the mirror is warped, and the problems it reflects are magnified and made monstrously grotesque. Human treatment of the environment has grown worse, and in our generation it has brought us to a point of crisis. Salinization, erosion, denudation of watersheds, silting of valleys and estuaries, degradation of arid lands, depletion and pollution of water resources, abuse of wetlands, and excessive population pressure, all are now occurring more intensively and on an ever-larger scale. Added to the old problems are entirely new ones, unsuspected in centuries past. Included among these are the encroachment of urban, industrial, transportation, and even recreational activities upon the landscape; as well as the appearance of nondegradable synthetics, pesticide and fertilizer residues, toxic and pathogenic wastes, radioactive isotopes, contaminated soil and groundwater, eutrophic rivers and lakes, smog and acid rain, mass extinction of species, and finally, the threat of global climate change.

Among the many nations abusing their natural endowment, America is not the least offender. This country's fundamental strength depends on its great soil and water resources, and their wasteful and destructive exploitation is surely sapping the nation's vitality and jeopardizing its future.

We can take no comfort at all in the fact that the problem is universal. Absurdly, nations fight wars over every inch of their political boundaries, while mindlessly sacrificing whole regions to environmental degradation. Their patriots salute the flag and take up arms to defend their country against external enemies, while neglecting its environment and ignoring the attacks being waged from within on the land they purport to love. Not enough people understand that, on the time-scale by which human individuals and societies typically reckon, the soil is a nonrenewable resource. So is a forest ecosystem, a river, a lake, or an aquifer. They belong not only to those who are the titled owners at any moment, but to future generations as well. In an even more profound sense, both soil and water belong to the biosphere, to the order of nature, and as one species among many, humans have no right to use them up.

The global impact of human activities on the biosphere can be considered in terms of the fraction of net primary productivity (NPP) humans have appropriated. Defined as the amount of solar energy fixed biologically by photosynthetic plants (the primary producers in any ecosystem) minus the amount of energy consumed by respiration, NPP provides the basis for the maintenance and growth of all consumers and decomposers. As such, it is the total food resource on earth. The more of NPP taken up by humans, the less remains for other species. There are various estimates of how much of NPP is used by humans. According to some estimates, however, humanity is already using close to 40% of terrestrial NPP. This implies that we are only one doubling away from usurping the greater part of the total biotic activity of our planet's land surface. If human numbers and economies double just once more, as may well occur within the next half century, little will remain for natural forms of life not directly serving our needs or desires. We will then be living in a greatly impoverished world of our own making, in an *egosystem* rather than a true ecosystem (Hillel, 1992).

An awareness that human activities are now threatening the biological balance on a global scale does not require us to romanticize the primitive life, as did Jean-Jacques Rousseau, and to advocate a total renunciation of materialism and progress. There probably never was a golden age of perfect human happiness in a pristine world, which in any case could not have lasted in the face of the innate aptitude of humans for meddling with the environment, and their urge to enlarge their domain and multiply in number. Nor, on the other hand, need we accept the notion expressed by Thomas Hobbes that modern man is infinitely better off than the primitive child of nature whose life was "solitary, poor, nasty, brutish, and short." These diametrically opposed notions are equally simplistic. Clearly, however, something has gone wrong in our relation with nature, and we must seek ways to redress it.

We live in an age and culture that is very sensitive to human rights, but does not grant equal weight to human responsibilities. We insist on our prerogatives, and neglect our obligations. Our attitude toward the environment is marked by careless confidence and reckless self-indulgence. These are attitudes and actions that, in any individual, we recognize as childish. And just as a mature person must learn to consider the circumstances and needs of others, so a mature society must restrain its profligate exploitation of resources and careless disposal of wastes, and begin to consider both the rights of future generations and the needs of other species.

Our hurried and impatient society has practically no mechanisms for long-term consideration or control. Most of our institutions and decision makers are concerned primarily with the problems of the moment: businessmen with the quarterly or yearly bottom-line, politicians with the next election. Even the Supreme Court deals only with legal issues brought before it and can hardly anticipate or initiate action on issues that are not yet imminent. This state of affairs places an extra responsibility upon us as scientists. It challenges and tests our ability to respond not only to current problems, but also to foretell how processes operating today may generate the problems of tomorrow. Such an undertaking propels us from the realm of certainty in which we feel secure into the uncomfortable realm of uncertain likelihood. An example of this situation is the issue generated by the enhanced greenhouse effect and the possibility of global warming. What shall we do when our crystal ball is murky? The obvious answer is to ask for more funding and time to conduct more research. But what in the meanwhile? Should we avoid any opinion or advice to the public and to policy makers begging for guidance, or should we risk committing error and offer an opinion (tempered, of course, by all due caveats) despite the uncertainty? Wherein lies the greater risk, in taking a position or in remaining aloof of controversial issues?

There is an old adage concerning the difference between the clever and the wise: The clever are those who are adept at extricating themselves from situations that the wise would have avoided from the outset. Unfortunately, our cleverness as a species has gotten us into an environmental dilemma from which the same faculty can no longer free us. Cleverness has reached and exceeded its limits. Wisdom is now needed.

The wisdom we need will not be found ready-made in any single profession or organization. It can only be developed through interdisciplinary, interinstitutional, and international research and cooperation. And it must enlist the best and most concerned social scientists as well as environmental scientists and technologists. All of us must learn to communicate across arbitrary professional boundaries and jointly search for ways to improve the prospects for all the world's children, while protecting the environment.

Examples of the need for a more comprehensive approach can be seen everywhere. One example is the need to avoid the progressive degradation of marginal land in semiarid regions. Some of that land should be retired and allowed to regenerate as a natural habitat. But then, how to feed the people who now derive their precarious livelihood from such land? The obvious answer is to intensify production on the most suitable and stable lands. But how? No single prescription is likely to provide a panacea in a complex environment. The most likely approach, and it must be a progressively developing approach, will depend on an optimal combination of improved practices that is best adapted to the physical, biological, and human circumstances in each case.

The history of humanity has always been a race between learning and disaster. The form of disaster threatening human welfare seems to change from time to time like a many-headed dragon. Hence the dragon cannot be slain once-and-for-all. Needed is an unrelenting effort on the part of scientists, in alliance with concerned citizens of all professions, to achieve a higher level of knowledge and understanding as a guide to more effective action. All this may require a redirection of science from a sectarian and individualistic mode to a more cooperative and inclusive mode. And it will require a reconsideration of our self-image and role as a species. We cannot hold to the anachronistic and arrogant assumption that our species is superior and free of the laws of evolution and ecology. Just as Copernicus informed us that our planet is not the center of the universe; just as Darwin included us in the animal kingdom; and just as Freud revealed that we are not even in control of ourselves; so the new environmental awareness should cure us of the pretense that we are (or even that we ought to be) in total control of nature.

There are hopeful signs that such a coalition is forming. A growing environmental movement is already playing an important role in many countries and international institutions. This movement cries for guidance in order that it might overcome inertial resistance or complacency while avoiding exaggerated alarmism. Public opinion can be awakened and mobilized on behalf of responsible environmental causes both locally and on a global scale. Though the proliferation of organizations and the din of numerous conferences, each vying for attention, may seem excessive at times, their underlying purposes are serious and positive.

Lately, even politicians have begun to express concern for the environment. The excellent work of the World Commission on Environment and Development (chaired by Gro Harlem Brundtland of Norway), embodied in the publication *Our Common Future* (Brundtland, 1987), is a significant milestone. Another landmark achievement is the 1987 Montreal Convention for the protection of the stratospheric ozone layer by limiting the manufacturing and use of chlorofluorocarbons (CFCs). The environmental conference held in Rio de Janeiro in June 1992 enlisted the participation of 120 heads of government, more than ever assembled in any other cause. Religious denominations of varied faiths have come to recognize and to address the environment as a common moral issue. Governments in the developing countries, initially resentful of environmental preachings from the wealthy countries, have also come to realize that their most vital national interests are indeed at stake. And there are favorable demographic trends: the rate of population increase has begun to decline. An auspicious start has been made, but it is only a start.

It seems to me that the proper progress of civilization is not only material, but conceptual and spiritual as well. It consists of concentrically widening spheres of inclusion. The earliest societies were characterized by an atavistic, narrowly exclusive view of the world: humans banded in small groups that regarded all others as potential or actual enemies. Gradually, that view has evolved as human notions of kinship were extended beyond an immediate clan to include progressively larger associations such as tribe, village, city, country, nation, creed, and, eventually, all of humanity.

This expanding perception of kinship and allegiance should now be consciously expanded even further to include the entire biosphere and the totality of life on earth. If such a transcendental notion is to become more than a nebulous mystical ideal, it must be rooted in fundamental, holistic science.

The timeless association of humanity with Nature is exemplified in the name Adam, the Bible's first human earthling, derived from the Hebrew noun *adama*, signifying soil, or earth. Likewise, the name of Adam's mate, Hava, (rendered Eve in transliteration) literally means "living." Together, therefore, Adam and Eve signify "Soil and Life." The same notion is echoed in the Latin name for man, *homo*, derived from *humus*, the stuff of life in the soil. Since the words humility and humble also derive from humus, it is rather ironic that we should have assigned our species so arrogant a name as *Homo sapiens sapiens* (wise wise man). As we ponder our past and future relation to the earth, we might consider changing our name to a more modest *Homo sapiens curans*, with the word *curans* denoting caring or caretaking, as in curator. Having failed to deserve the old name, we will need to work all the harder to deserve the new one.

REFERENCES

Brundtland, G.H. (chair). 1987. Our common future: Report of the World Commission on the environment and development. Oxford Univ. Press, Oxford.

Hillel, D. 1992. Out of the earth: Civilization and the life of the soil. Univ. of California Press, Berkeley.

2 Soil and Water as Critical Factors in the History of the Fertile Crescent

Robert McC. Adams
Smithsonian Institution
Washington, DC

The Fertile Crescent has reference to the great arc of early agricultural life and city development in the Near East, anchored at its two ends by the fertile river valleys of the Nile and the Tigris and Euphrates Rivers. Since it was largely semiarid to arid, a concentration on the deployment of water resources was one of its most pervasive characteristics. Taking this as central, I concentrate here not on the broad intermediate area, but on the great irrigation regimes at the two ends of the crescent.

The Nile is and always was, relatively the more benign, stable, and reliable of these three river systems from the viewpoint of human use. Its position is substantially fixed within a narrow valley bounded by desert scarps. While the Nile's flow is variable, it falls mostly within moderate limits of variation that only infrequently created serious high- or low-water hardships. Egyptian agriculture, given insignificant rainfall is completely dependent on irrigation. There are suggestions in some early art of a politico-religious character that swamp drainage may have been undertaken as a royal initiative at the very beginning of the Old Kingdom. But as remains the case in later tomb paintings, Egyptian art offers idealized, ideological statements that may not closely reflect reality. Still, it is surely not unlikely that the Nile Valley, at the time of its first extensive cultivation in the fifth millennium BC (cattle may have been herded in the then-wetter Sahara much earlier) was a densely vegetated wilderness.

Irrigation in the Nile Valley followed a comparatively simple basin regime that required large-scale canalization only in a few localities like the Fayyum depression (Butzer, 1976). Simple, hand-operated lifting devices were in widespread use; geared, animal-operated ones followed only much later and more rarely. Given the limited canalization, silt clearance does not seem to have been a substantial part of the agricultural cycle. Soil salinization was similarly limited, since the Nile's annual flood provided a flushing action, except (as now) in the lower delta fringes along the Mediterranean. While my own knowledge of ancient Egypt is based largely on secondary sources, I have never encountered a reference to any textually expressed concern over salinization and how to deal with it. Not only a water source, the Nile itself was of course the overwhelmingly dominant

Soil and Water Science: Key to Understanding Our Global Environment, SSSA Special Publication 41.

artery for communication and commerce. Canal offtakes from it may have been primarily devoted to facilitating the construction of monumental tombs and servicing the temple complexes attached to them.

Obviously, Egyptian civilization could not have existed without its annually replenished, highly fertile soil and the Nile's plentiful water. But the relationship has to be seen as sustaining, not as a dynamic one. There was enormous conservatism and continuity in the technology and methods employed in the irrigation system.

The Tigris and Euphrates present another story altogether. Note first that it is somewhat misleading to couple or hyphenate them. Neither river was available for irrigation above the Mesopotamian alluvium (beginning not far north of the latitude of Baghdad) before modern times, except in minor local areas and along some tributaries. The uplands were a zone (if with frequent failures) of dry farming that was nevertheless productive enough, together with pastoralism, to sustain cities.

In the alluvium, the Tigris was too powerful and dangerous, and probably for at least the upper part of the alluvium too deeply entrenched, to be used for irrigation at all until quite late (mid-first millennium AD). Moreover, its ancient course or courses are hard to locate. It now appears that at least some branches of the Tigris and Euphrates may have been periodically connected in antiquity (naturally? artificially?). But in any case, there were few settlements of any substantial size that adjoined or depended upon the Tigris until very late in the pre-Christian era. Its adjoining region along both banks may have served principally as a zone reserved for herding.

Turning to the Euphrates, its flow was sufficiently more modest than that of the Tigris to be more manageable. But annual variability was comparably high, so that irrigation agriculture was always vulnerable. The major ancient courses of the Euphrates, well-defined by adjoining city-ruins and traces of ancient meander-belt levees, lay down center of alluvium and not in their present position near its western edge (Adams, 1981). Rainfall, while very low, could occasionally be a significant supplement since it seldom coincided with variations in river amplitude.

Annual variations were only part of the problem. There was also grave uncertainty that persisted throughout the main winter growing season as to whether volume might increase or decrease sharply without significant advance warning. This intensified the tendency to over-water when supplies were available, thus exacerbating upstream–downstream conflict. Moreover, attempts to protect prospects for the next harvest by applying more water than was minimally necessary were an inevitable source of long-term problems with salinization, since their effect was to raise the level of highly saline groundwater to root levels.

The scale of irrigation of the ancient Mesopotamian plain was variable, linked to political centralization, and stability. Truly gigantic systems did not precede the Christian era, and were linked to a corresponding growth of population. Desilting of secondary and tertiary canals, especially near offtakes, became a major requirement as the landscape was comprehensively altered in this essentially artificial, nonequilibrium direction. After a large-scale abandonment in the 10th century AD or so, this latticework of abandoned levees was subjected to aeolian erosion. The result was the appearance of great belts of dunes in uncultivated areas that, as they have moved continuously, have eroded away considerable depths of overlying alluvial deposits. Exposed in this way have been substantial land-surfaces as they were in the third and even fourth millennium BC.

A "barren mockery of snow," salinity was an omnipresent problem. Until very recently, it was understood only as a destructive surface-salt infestation that might be corrected by physical removal. That ineffective process, carried on under barbarous conditions by great gangs of slaves of largely East African origin in the ninth century AD, led to a prolonged rebellion that nearly brought down the Abbasid Caliphate.

There are sophisticated traditional techniques for, e.g., melon growing, that reduce the effects of salinity and may well be ancient. But there was no appreciation of deep drains, which are now playing a very significant role, or apparently even of the dangers of over-irrigation. Detailed, carefully recorded observations were kept during long periods of salt-affected portions of field that were no longer cultivated for this reason. There was a shift away from wheat (*Triticum* sp.) toward an overwhelming concentration on more salt-tolerant barley (*Hordeum vulgare* L.) that was already underway by early in third millennium BC, but this also may reflect the development of a wool textile export industry to permit long-distance trade, with the barley a preferred fodder for sheep (*Ovis aries*). On any reconstruction, we assume that land was simultaneously being abandoned because of salinization and eventually reclaimed during most of antiquity. That is still the case today.

Modest irrigation canals have been identified in upland tributaries that were in operation already in the sixth millennium BC. Canal rather than basin irrigation was clearly in widespread, if diffuse, use in the alluvium not long afterward. New cuneiform terminology attests significantly advanced technology of weirs, reservoirs, and specialized control gates by no later than the early third millennium BC. There are records of labor parties for large-scale brick dam construction by the end of that millennium, but modern observations indicate that a lot of this can go on under strictly local initiative. The *state*, briefly to mention an old theoretical controversy, was apparently not a consequence of large-scale irrigation, but certainly a promoter of some of the more grandiose developments of it. By Islamic times, some of these developments, like the Nahrwan Canal system that extended across the entire fan of the lower Diyala River, were grandiose indeed (Adams, 1965; El-Samarraie, 1972).

I must emphasize that this was not mono-crop agriculture, but a complex subsistence system with institutions for exchange, marketing, rationing, and large-scale storage (especially grain storage for rations). Orchards and garden crops were concentrated along the levee crests, nearest to reliably perennial water sources. Cereals were grown in fields on levee back-slopes, while sheep and goats (*Capra hircus*) were herded in the seasonally watered depressions and on the more remote dry steppes. Cattle (*Bos* sp.) and pigs (*Sus scrofa domesticus*) both played a significant role in ancient diets, and tended to be kept closer to the settlements and probably often in the marshes. Fish were a vital part of the diet also.

Soil fertility is still an unsolved problem. Incredibly high yields, given as multiples of seed inputs, are extensively attested in the texts. How do we explain them? Partly, they may reflect very sparse seeding, which would help to explain the prevailingly very low price of land. If sparse seeding led to sparse watering, this might have been a salt control measure, whether conscious and deliberate or not (Postgate, 1988–1990).

All in all, on the Mesopotamian alluvium both soil and water presented harsh challenges, not blessings. The combination could yield abundant subsistence resources. It may even have served as a primary stimulant to the development of social hierarchies and complexity, but only with much work and the overcoming of grave uncertainties.

REFERENCES

Adams, R. McC. 1965. Land behind Baghdad: A history of settlement on the Diyala plains. Univ. of Chicago Press, Chicago.

Adams, R.McC. 1981. Heartland of cities: Surveys of ancient settlement and land use on the central floodplain of the Euphrates. Univ. of Chicago Press, Chicago.

Butzer, K.W. 1976. Early hydraulic civilization in Egypt: A study in cultural ecology. Univ. of Chicago Press, Chicago.

El-Samarraie, H.Q. 1972. Agriculture in Iraq during the 3rd century AH. Librarie du Liban, Beirut.

Postgate, J.N. (ed.) 1988–1990. Irrigation and cultivation in Mesopotamia. Part 1 and 2. Bulletin on Sumerian Agriculture. Vol. 4 and 5. Cambridge Univ., Cambridge.

3 The Future Role of Irrigation in Meeting the World's Food Supply[1]

Montague Yudelman

World Wildlife Fund
Washington, DC

RECENT TRENDS IN FOOD PRODUCTION AND FUTURE DEMAND

The post World War II era has been one of remarkable growth in agriculture and food production in the world at large and more particularly in the developing countries. One indicator of the sustained increase in the supply of food has been the steady and consistent fall in grain prices; the real price of rice (*Oryza sativa* L.), the most widely grown and consumed cereal in the developing countries, has fallen from $600 a ton in 1950 to $200 a ton in 1991 and the price of wheat (*Triticum aestivum* L.) has fallen from $300 a ton to $190 a ton during the same time period (World Bank, 1992b).

A second and more persuasive indicator of the increase in food supply is that the estimated food availability per person in the world has increased by 17.5% between 1960 and 1990 and by 27.6% per person in the developing countries. The largest increases have been in China (55.5%) and the lowest in Africa (10.5%). During the past 30 yr, despite a very rapid increase in population, food supply has increased to the point where there is now food enough in the developing countries as a whole, so that every individual could have an adequate diet to meet his or her basic needs. This is a significant achievement even though there are still shortfalls in supply in limited parts of Africa and there continues to be a large amount of poverty induced malnutrition in the world (Anderson, 1992).

The very substantial increase in food supplies have been accompanied by a significant change in the source of agricultural output. Prior to 1960 most of the increase in agricultural production in the developing countries came from an expansion of acreage either through bringing fallow land into cultivation or from expanding the frontiers of production into forests and grasslands. In addition, in the past, most farmers used very little in the way of purchased inputs from off the farm. They relied largely on family labor and land extensive practices of shifting cultivation to maintain the fertility of their lands. Between 1960 and 1990, though

[1]This is a condensed extract by Yudelman (1992). This publication details the references used as a basis for this article.

Soil and Water Science: Key to Understanding Our Global Environment, SSSA Special Publication 41.

there have been dramatic changes. This is clearly illustrated by cereal production in the developing countries, which grew by 118% between 1960 and 1990; 92% of that increase came from higher yields per hectare and only 8% from expanded acreage. The largest shifts were in the land scarce economies of Asia, the smallest, in land abundant Latin America and sub-Saharan Africa. Even so, more than one-half the increase in output in sub-Saharan Africa came from increased yields (World Bank, 1992a).

The transformation from extensive to intensive agriculture has been attributed to many technical, social, and economic factors. There is widespread agreement though, that three elements contributed substantially to this transformation. These are:

1. The development and diffusion of modern high yielding varieties of seeds [especially rice, wheat, and corn (*Zea mays* L.)] that give substantially higher yields than traditional varieties when planted with appropriate inputs of fertilizer and given regular supplies of water. Almost 80% of the wheat and rice areas in the developing countries are now planted with modern varieties.
2. The development of low cost industrial methods of manufacturing nitrogenous fertilizer and its widespread diffusion and adoption by farmers, especially in Asia. Between 1970 and 1990, fertilizer consumption in the developing countries has increased from 256 kg ha^{-1} of available land to 853 kg ha^{-1}. The average levels of use are highest in Asia, but are still low compared with those in Europe; even so the increases in use have contributed in a large way to the growth in yields that have taken place in the developing countries (World Resources Institute, 1992).
3. The very large increases in investment in irrigation in the developing countries has led to the acreage under irrigation growing from 100 million hectares in 1960 to 174 million hectares in 1990 (author's estimate). Nearly one-half of the increase in irrigated acreage has been in semiarid areas and has enabled farmers to grow crops where very little grew before. The other one-half, in the humid tropics, has supplemented rain fed agriculture, so enabling farmers to grow more than one crop per year even when rains were tardy (Technical Advisory Committee, 1992).

The combination of the introduction and diffusion of improved varieties, the spread of fertilizers and the extended control of water supplies has led to substantial increases in yields. Between 1960 and 1990 yields of staples (cereals) rose by a 117% in Asia, 70% in Latin America, and 45% in Africa. Overall the annual average increase in yields of cereals during the last three decades has been 2.6% a year, a rate higher than that of population growth (World Bank, 1992a).

THE ROLE OF IRRIGATION

The irrigated areas have made an impressive contribution to food production. According to Food and Agriculture Organization (1992), there were close to 868 million hectares of arable land in use in the developing countries as a whole in

1989 to 1990; 173 million hectares, or 19.9% of all arable land was irrigated. The distribution of this irrigated acreage is heavily concentrated with 131.7 million hectares, or 78.2% of all irrigated land in the developing countries being in Asia; the Middle East and North Africa have 18.6 million hectares, or 10.7% of the irrigated area, Latin America and the Caribbean have 14.07 million hectares, and 8.1% of irrigated area, while sub-Saharan Africa has 5 million hectares and only 3% of the irrigated land in the developing countries. Three countries in Asia: China with 45 million hectares of irrigated land, India with 43 million hectares of irrigated land, and Pakistan with 16 million hectares of irrigated land, account for two-thirds of all irrigated land in the developing countries. The next three most important countries in terms of irrigated acreage are Indonesia (7.3 million hectares), Iran (5 million hectares), and Mexico (5.3 million hectares). The remainder of the countries have < 5 million hectares of irrigated land each. No country in sub-Saharan Africa has >1 million hectares under irrigation (Food and Agriculture Organization, 1992).

One measure of the importance of irrigation as a factor in food production is the share of food production that comes off irrigated land. The Technical Advisory Committee (TAC) of the Consultative Group for International Agricultural Research (CGIAR), acknowledging very thorny problems of valuing both traded and nontraded commodities, has estimated that between 1987 and 1989, the annual value of all crop production in the developing countries was in the neighborhood of $364 billion dollars; it is estimated (by the author) that $104 billion dollars worth of crops, or 28.5% of the value of all crop production, was produced on irrigated land. More than 30% of all food production, valued at around $96 billion, however, was grown under irrigation. Perhaps irrigation's largest contribution to both consumers and producers is that an estimated 46.5% of all grain and 57% of the total value of the most widely grown basic staples, rice, and wheat, were produced under irrigation (Technical Advisory Committee, 1992).

Irrigation's contribution to production would best be measured by isolating the contribution of controlled supplies of water to output. The contribution of each input such as water and fertilizers, however, cannot be measured by simply considering the difference in output with and without each input because of the strong interaction among the inputs. The impact of the joint effect of the inputs on output can only be estimated under carefully controlled multifactor experiments. Such an experiment was undertaken by IRRI in the early 1970s. The experiment sought to explain the differences in yields of rice obtained at the farm level compared with the attainable potential as demonstrated on research stations. A conclusion was that the lack of control over water was the single biggest constraint. If all rice was fully and properly irrigated, maximum yields could average 5.6 tons per hectare so that water control was responsible for 23% of the difference between maximum possible and actual yields (Herdt & Wickham, 1978).

On a regional basis, it is estimated that ≈60% of the value of crop production in Asia is grown on irrigated land. This includes ≈80% of Pakistan's food, 70% of China's and >50% of India's and Indonesia's food. In the Middle East and North Africa, more than one-third of the region's crop production by value is irrigated, including all the food grown in Egypt and more than half of that grown in Iraq and Iran. A relatively small proportion of agricultural production in Latin America, ≈10%, is grown under irrigation, but more than one-half of all the food production of crops grown for export in Chile and Peru are irrigated. Sub-Saharan Africa, with the smallest regional acreage under irrigation, produces estimated 9% of its total food production on irrigated land. Madagascar produces >20% of its agricultural output and food from irrigated land (author's estimate based on Food and Agriculture Organization data, Food and Agriculture Organization, 1992).

The irrigated sector has performed an essential task in meeting the basic food needs of billions of people in the developing countries, especially in Asia. In the past it has provided more than one-half of the two most important basic staples and close to a one-third of all food crops. In the future, given the importance of increasing yields, the irrigated sector will have to provide an even larger proportion of total food output, especially in Asia, which depends so heavily on irrigated agriculture.

FUTURE DEMAND

Future demand for food will depend on a host of factors including the rate of increase in population, increases in income and the proportion of any increase in income that consumers will spend on food. The most important determinant of growth in demand will continue to be population growth. It is widely accepted that a demographic transition is in process in most of the developing countries and that population growth is slowing down in all of the world other than in sub-Saharan Africa and the Middle East. At the same time, incomes are expected to rise substantially during the next 35 to 50 yr (apart from sub-Saharan Africa). When these factors are taken into account along with assumptions about changes in tastes as income rises, it is projected that the rate of increase in demand for food will fall from ≈3.1% a year to slightly >2% by the year 2025 and by ≈1.5% by 2050. (The decline in population growth from 2% to ≈1.13% up to 2025 then to <1% being the most important determinant of the fall in demand up to 2025 and 2050.) In sub-Saharan Africa and the Middle East, because of the continued high growth of population, demand will continue to grow by ≈3% a year well into the next century before slowing down. While overall growth rates are slowing down, there will still be an increase of between 80 to 90 million people a year with the population of Asia being projected to rise to >5 billion (or more than the present global population) and that of Africa to by >1 billion by 2025 and substantially more than that by 2050 (Yudelman, 1993).

Food production in the developing countries has been rising by close to 3% a year. Consequently, if recent trends are sustained there should be every prospect for meeting further growth in demand of between 2 and 3% a year by 2025. Most of the increase in supply of basic foods, such as cereals, would have to continue coming from increased yields. Thus in landscarce Asia, which is the producer and consumer of most of the world's rice, and where there has been no expansion of acreage under rice during the past 20 yr, yields will have to rise by an average of 2% a year to meet projected demand. By 2025 average yields throughout Asia will have to double, or be at same levels as current yields in Japan with its highly intensive agriculture.

IRRIGATION AND FUTURE STRATEGY FOR RAISING YIELDS

There has been a noticeable slowing down in the rate of expansion of irrigation in the 1980s. Part of the reason for this is that the real capital costs of new irrigation have risen on average, by between 70 and 116% per hectare, (rising costs indicate a growing shortage of well sited, low cost options for development of irrigation; Rosengrant, 1992). In addition, dam construction has been curtailed because of widespread complaints by environmentalists and because evaluations of past investments have highlighted very low levels of efficiency of operations and poor maintenance of existing irrigation systems. Even though irrigated agriculture has led to increased output, most irrigation systems are operating well below their capacity and are giving much lower increases in agricultural output than planned. These and other concerns have led the major international investors and donors to slow down their commitments to expanding irrigation and to focus on rehabilitating existing systems. The point has now been reached where there is active debate about whether the extent of the spread of irrigation-induced salinition is exceeding additional acreage brought under irrigation so leading to an annual loss in area irrigated.

There is also reason to question the sustainability of past rates of expansion of irrigation. There have been several estimates of the physical potential for expanding irrigation. Most of these estimates are based on criteria that center around agroclimatic conditions where controlling water would lead to increased output from productive, but underutilized soils. The estimates that have been made suffer from the limitations of available data. There are wide differences in some of these estimates, e.g., Food and Agriculture Organization estimated that there were 3.5 million hectares of land available for irrigation in Zambia, while the World Bank estimates that the potential was only 12% of that amount or 420 000 hectares (Olivares, 1987). Most of the estimates, though, agree that there is scope for expansion (though some countries, in the Middle East and North Africa, are beginning to reach the outer limits of expansion from available water supply).

One comprehensive estimate of the overall potential for expansion in the developing countries prepared by the World Bank and the UNDP concludes that there is scope of 59% increase in acreage under irrigation in the developing countries as a whole. The largest potential for increase (69 million hectares) is in Asia, especially in India and China. This is followed by South America with 20 million hectares with most of this being in Brazil. Sub-Saharan Africa has a potential to increase irrigated acreage by >470%, from a reported 3.4 million hectares to 16.5 million hectares, the largest potential increase being in Angola. The most limited opportunities for expansion are in the Middle East, Central America and North Africa (Crosson & Anderson, 1992).

On the face of it there would appear to be considerable scope for expanding irrigation. If past rates of increase in the expansion of irrigation are continued, however, then the prospects are that the available potential would be exhausted well before 2050. This is most notable in the case of Asia, the region that is most dependent on irrigation for its food supply. Between 1960 and 1990, acreage under irrigation grew from 87 million hectares to 147 million hectares, or at a rate of 1.82% a year; the rate of growth slowed down between 1980 and 1990 when the acreage under irrigation rose by 1.26% a year from 129 million hectares to 147 million hectares. The World Bank and UNDP estimates that the potential exists in Asia, to irrigate 228 million hectares. Consequently, if irrigation was to expand at the same pace as it has during the past 30 yr then all irrigation potential would be exhausted by 2015; if it was to expand at the much slower rate of the last decade then the potential would be exploited by 2025.

The prospects of a limitation on the expansion of irrigation in Asia is disconcerting as the increase in irrigation in that region was one of the engines of agricultural growth. It is probable that rising costs will slow down irrigation expansion well before the limits are reached. This will raise the premium on improving the efficiency of all irrigated agriculture to meet future demand, i.e., increasing yields on the 150 million hectares already irrigated will probably give greater returns than high cost development of additional acreage under irrigation. Given the high cost of developing new acreage, every effort will be necessary to make all irrigated acreage more productive than in the past.

CONCLUSION

Irrigation will continue to be of major importance in providing future food supplies for the growing population in the developing countries, especially in Asia. Rising real costs of developing irrigation and the slowing down of the expansion of the area under irrigation, however, are harbingers of the approaching limits to the low cost expansion of irrigation in the next century. Consequently, there will have to be less reliance on new irrigation and more emphasis on using all available water more effectively than at present to increase yields to meet future demand. This will require changes such as the introduction and adoption of water charges that reflect the true value of water, the incorporation of drainage into irrigation systems, the investment of funds for maintenance rather than the more politically appealing notion of expansion, the participation of water users in management and closer cooperation between managers of irrigation systems and farmers. Improving the management and design of irrigation systems per se can save water and reduce losses of land from water logging and salination. This will permit larger areas to be irrigated and to increase output. But if irrigated agriculture is to make a full contribution to increasing food production, this will only be brought about if there are also improvements in other facets of agricultural development: improvements in such areas as price policies, plant breeding and programs to make plant nutrients and other inputs available to the hundreds of million of farmers who produce food in the developing countries. Thus the future role of irrigation in meeting the world food supply will depend not only on engineers but also on plant breeders, soil scientists and agronomists, along with pragmatic scientists such as our honoree, today, whose career has shown he possesses many of the attributes that will be needed to help make irrigated agriculture fully productive in the future.

REFERENCES

Anderson, P.P. 1992. Global perspectives for food production and consumption. Int. Food Policy Res. Inst., Washington, DC.

Crosson, P., and J.R. Anderson. 1992. Resources and global food prospects: Supply and demand for cereals to 2030. World Bank, Washington, DC.

Food and Agriculture Organization. 1992. World food survey. FAO, Rome.

Herdt, R.W., and T. Wickham. 1978. Exploring the gap between potential and actual rice yields: The Philippine case. p. 74–78. *In* Economic consequences of the new rice technology. IRRI, Los Banos, Philippines.

Olivares, J. 1987. Options and investment priorities in irrigation development. World Bank/United Nations Development Program. World Bank, Washington, DC.

Rosengrant, C. 1992. Sustaining rice productivity growth in Asia. A policy perspective. Int. Food Policy Res. Inst., Washington, DC.

Technical Advisory Committee (TAC). 1992. CGIAR priorities and strategies. Part 1. Rome, Italy.

World Bank. 1992a. World development report 1992. World Bank, Washington, DC.

World Bank. 1992b. Market outlook for primary commodities. Vol. II. World Bank, Washington, DC.

World Resources Institute. 1992. World resources report. Basic Books, New York.

Yudelman, M. 1993. Demand and supply of foodstuffs up to 2050 with special reference to irrigation. Int. Inst. for Management of Irrigation, Colombo, Sri Lanka.

4 Is Irrigated Agriculture Sustainable?

J. Letey
University of California
Riverside, California

Irrigation is human manipulation of the natural hydrologic cycle for the purpose of increasing crop production which, on the average, has been highly successful. More than one-third of the total harvest has been produced on irrigated land, which amounts to only ≈17% of the world's cropland (Hillel, 1991, p. 227). Clearly, irrigated agriculture serves a significant role in producing food and fiber for which demand is expected to increase with increasing human population. The question "Is irrigated agriculture sustainable?" is a critical question for the future of humanity.

The concept of sustainability is ambiguous. A dictionary definition of *sustain* is "to keep in existence". The failure to keep irrigated agriculture in existence at some level would appear to be a preposterous thought. Yet, Mesopotamia provides an ominous historical precedent. Hillel (1991) documents the rise and fall of this great society. The rise was associated with surplus production by farmers largely aided by irrigation developments that freed much of the population to pursue other professions associated with urban societies. Declining yields brought about by waterlogging or salinization were devastating to cities where the needs of a considerable superstructure of priests, administrators, merchants, soldiers, and craftsmen had to be met by surpluses from agricultural production. The southern part of the alluvial plain never recovered and many great cities dwindled into villages or were left in ruins.

A political definition of sustainable agriculture (Section 1404 of the Natural Agricultural Research, Extension and Teaching Policy Act of 1977, as amended by Section 1603 of the FACT Act) is "an integrated system of plant and animal production practices having a site-specific application that will, over the long term: (i) satisfy human food and fiber needs; (ii) enhance environmental quality and the natural resource base upon which the agricultural economy depends; (iii) make the most efficient use of non-renewable resources and on-farm resources and incorporate, where appropriate, the natural biological cycles and controls, (iv) sustain the economic viability of farm operations; and (v) enhance the quality of life for farmers and society as a whole." I find this definition too complex and cumbersome to address in my evaluation of the sustainability of irrigated agriculture.

Soil and Water Science: Key to Understanding Our Global Environment, SSSA Special Publication 41.

This chapter will consider whether the present level of crop production associated with irrigated agriculture can be infinitely maintained or increased, and if so, under what conditions and at what costs can the production level be maintained? The level of crop production under irrigation is constrained by: (i) physical–biological (quantity of water, quality of water, and quality of land); (ii) societal; and (iii) economic factors. The sustainability of irrigated agriculture will be evaluated in the context of these constraints. Emphasis will be placed on crop production and not merely the spreading of water.

QUANTITY OF WATER

Water balance for a geographic area is schematically presented in Fig. 4–1. Inputs include precipitation, and surface and subsurface inflows. Outputs consist of evapotranspiration (ET), and surface and subsurface outflows. Storage can exist at the surface in reservoirs or in the subsurface in geologic strata. Water contained in coarse-textured geologic strata from which water can be economically extracted by inserting a well is considered to be retrievable with a potential for human use. Water stored in other geologic strata is nonretrievable and has no direct human utility.

The long-term sustainable level of irrigation is constrained by the amount of precipitation and to the extent that the other components of the system can be manipulated. At steady state (storage is constant):

$$\text{Crop ET} = \text{precipitation} + \text{surface inflow} + \text{subsurface inflow}$$

$$- \text{surface outflow} - \text{subsurface outflow} - \text{noncrop ET}$$

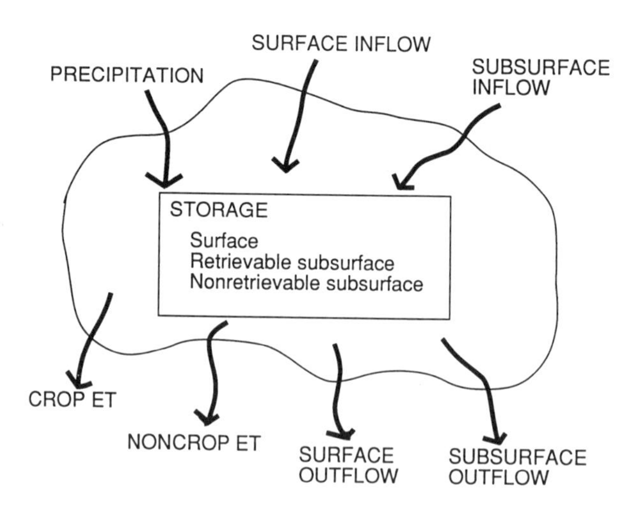

Fig. 4–1. Components of water balance in a geographic area.

Care must be taken to use comparable units in computing a water balance. The surface flows are usually expressed as volume of flow passing a point per unit time. Subsurface flows are usually expressed as Darcian flow rate with dimensions of length per unit time. Precipitation and ET are expressed in units of length per time, but are not usually expressed as instantaneous values. They may be expressed as a cumulative depth over an extended period of time such as a year. The storage quantity is usually expressed as a volume.

Inflow and outflow values are responsive to management within and surrounding the geographic region under consideration. Surface flows are most amenable to human manipulation and the effects are readily quantifiable. For example, rivers can be dammed and the rate of release through the dam can be controlled and measured. On the other hand, subsurface flows are very difficult to quantify and are almost always impossible to control. Subsurface flows are the unobserved (and frequently ignored) consequence of manipulating surface flows. Building dams, pumping water, and irrigating lands all alter the hydraulic gradients, inducing a change in direction or rate of subsurface flow.

Precipitation is highly variable, both within and between years. Thus, the steady-state value must represent the average for several years and the storage capacity must be large enough to smooth out the temporal variability. In principal, steady-state irrigation could be infinitely perpetuated at the level consistent with the management decisions affecting the inflows and outflows, except for constraints to be specified later.

In some instances, irrigation is being carried out in excess of the steady state value by exploiting subsurface storage at a more rapid rate than it is being recharged. Hillel (1991) reports several examples of this situation. One of the most prominent cases is in the USA where the Ogallala Aquifer in the Great Plains region is being exploited. In parts of Kansas, Colorado, Oklahoma, New Mexico, and Texas, water has been withdrawn far in excess of natural recharge. As a consequence, the water table is lowered and irrigation at this level cannot be infinitely sustained. As the water table drops, the cost for pumping water increases, and the demand for water diminishes. As typical for most exploited resources, the rate of exploitation rapidly increases, reaches a maximum level and then gradually decreases. Ultimately, the irrigation level will be limited to the aquifer recharge rate.

WATER QUALITY

Water quality encompasses the chemical, physical or biological properties of water that affect its use. In the context of this chapter, water use is for growing crops. The discussion in this section will be restricted to components of water quality that directly affect crop growth and the effective use of water for irrigation. Components of water quality that have little effect on crop growth, but have significant indirect effects on the level of irrigation will be reviewed in another section. Total dissolved solids (salinity), is one of the constraints on the steady-state level of irrigation. Salinity decreases the effective quantity of water, i.e., more saline water is required than nonsaline water to achieve the same crop production. No unique function exists that quantitatively relates effective water quantity to water quality. The relationship between these two factors is a function of crop, desired crop yield, and irrigation uniformity. Letey (1993) reported results of computed amounts of water of various salinities required to give maximum yield and 90% maximum yield of corn (*Zea mays* L.; salt-sensitive) and cotton (*Gossypium hirsutum* L.; salt-tolerant) when irrigated at different uniformities (Table 4–1). Irrigation uniformity was characterized by standard deviation (SD) with increasing SD representing decreasing irrigation uniformity.

As salinity of irrigation water (EC_i) increases, more water must be applied to produce the same yield. The increase is greater for the salt-sensitive corn than salt-tolerant cotton. The value is also dependent on whether maximum yield or less than maximum yield is desired. The net effect of salinity is to decrease the effective quantity of water, and the decrease is greater for salt-sensitive than salt-tolerant crops.

Precipitation is mostly free of dissolved salts, so one might assume that the sustainable supply of water is quantitatively related to the amount of precipitation. This assumption is compromised by management variables. For example, the salinity of the Colorado River increases as it winds its way from the Rocky Mountains through southwestern USA to Mexico. The increased salinity is greater than expected based on concentration of water by evaporation. The increase in salinity is largely attributed to flow of salty water into the river, in some cases related to irrigation. One such case occurs in the Grand Valley of Colorado. Water is diverted from the Colorado River for irrigation. Canal seepage and deep percolation below the root zone result in subsurface return flows to the river that passes through saline geologic deposits. So, even though the excess irrigation water returns to the river, it is saltier and its effective quantity for downstream use has been diminished. Alternatively, seepage can be restricted by lining the canals and irrigation could be managed to require less application and less deep percolation. This management would require less diversion from the river and also result in less return flow. Thus, the quantity in the river is not increased,

Table 4–1. The amount of applied water (cm) of various salinities (EC_i), necessary to achieve 100 or 90% yield of cotton or corn when irrigated at various levels of uniformity (SD).

	100% yield			90% yield		
		SD			SD	
EC_i	0.01	0.27	0.40	0.01	0.27	0.40
dSm^{-1}			Cotton			
0.1	73	105	>120	57	64	73
2.0	77	110	>120	61	68	78
4.0	90	>120	>120	66	75	85
8.0	†	†	†	82	95	105
11.0	†	†	†	110	120	>120
			Corn			
0.1	68	110	>120	61	69	78
1.0	83	>120	>120	69	78	89
4.0	†	†	†	†	†	†

†Yield cannot be achieved regardless of the amount of applied water.

but the water quality is increased. Consequently, the effective water quantity available for downstream users is increased. Whether the irrigation return flows migrate to rivers, add to the stored water in aquifers, or merely raise the water table in aquicludes and aquitards, the qualitative net effect is the same. The effective water supply is diminished. Consequently, the eventual steady-state irrigation level based on crop production is decreased.

The most easily observed detrimental effect of deep percolation is a rise of water table so it encroaches the root zone. Without an artificial drainage system to lower the water table, crop yields sharply decline and eventually the land gets "salted out." The historic example of Mesopotamia that was described by Hillel (1991) is the classic example of this phenomenon. If proper irrigation and design principles had been applied to this region, the military would have had to coin a different code name than "Desert Storm" for the invasion at Iraq.

The story of Mesopotamia is ancient, yet the story of Mesopotamia with minor variations has been repeated for millennia. Wilford R. Gardner's statement in reviewing the book by Hillel (1991) is particularly appropriate: "Dr. Hillel is to be especially commended for his sober reminder that what has happened before can happen again and again. Just because we know our history does not mean that we are not doomed to repeat it."

The story that has been historically repeated is development of irrigation projects without adequate consideration of drainage. A plausible explanation exists for this behavior. The benefits of increased crop production associated with irrigation projects are used to justify investment. Drainage facilities are not immediately required until the water table rises, so planning and investment for drainage are delayed. The requirement for drainage incrementally increases with the low-lying lands suffering first. Without a coordinated plan for both irrigation and drainage, installation of drainage systems may be physically difficult or very expensive. Since only a small land area is involved, investment may not be deemed justified. Investment in drainage at this time is only to maintain the present level of benefit and not enhance the benefits to a higher level. Means of disposing the drainage water must be considered as well as the physical drainage facilities. Options for drainage water disposal may be limited, and in some cases faced with unforeseen problems.

Development of irrigation in California illustrates the validity of the statement "Just because we know history does not mean that we are not doomed to repeat it." E.W. Hilgard, professor of Agriculture at the University of California, made a report to the president of the University of California in 1877 that was credited by Dorsey (1906) as being the first publication dealing with alkaline soil in the USA (as used by Hilgard the term *alkali* is synonymous with the present use of the term *salinity*). Hilgard was well-aware that development of salinity was a hazard associated with irrigation.

Although symptoms of salinity were being observed in some irrigated areas, it was not being publicized. Concerning early observations of salinity, Hilgard (1886) noted ". . . This question remained more a matter of curiosity than of serious concern. It was not until . . . the alkali was continually extending its area . . . where before there had been no sign of it, that the public interest was aroused; but, even then the fear of injuring the sales of land . . . caused many to avoid and frown down mention or discussion. But the specter would not be laid to rest by the hushing-up policy." He also recognized that subsurface drainage and leaching were required to maintain irrigated agriculture under saline conditions. Hilgard (1889) stated, "The time is not far distant when in California the laying of under-drains will be considered an excellent investment on any land as valuable as all irrigated land is likely to be; and when that day comes, alkali will be at an end on irrigated lands in this state."

Hilgard's optimistic prediction has not come to fruition. Indeed, the future of irrigated agriculture in the western San Joaquin Valley of California is very uncertain. The unfortunate state of affairs is partially a result of the repeated historically documented error of developing irrigation without adequate plans for drainage and partly due to unexpected problems associated with disposal of drainage water laden with toxic elements such as Se. The chronology of events related to irrigation and drainage in the western San Joaquin Valley published by Letey et al. (1986) will be summarized to illustrate the fragile nature of irrigated agriculture as it can be buffeted by political as well as physical-biological factors.

The Central Valley Project was first proposed in the 1930s as a massive California project to construct dams, canals, drains, and other structures on the Valley floor. The state had originally planned to construct the project, but during the Depression the state could not raise bond monies and the federal government assumed and assigned management responsibility to the U.S. Bureau of Reclamation. In the early 1950s, broader problems of basin-wide salt balance became apparent. A study in 1956 acknowledged that drainage systems would be necessary at lower elevations and that the most likely site for drainage water disposal would be the Delta, with discharge into the San Francisco Bay. Also in the 1950s, efforts were intensified to bring additional irrigation water supplies to the west side. Westlands Water District (WWD) was organized in 1952 upon petition of landowners of 162 000 ha in need for supplements to the rapidly depleting underground water supply. The San Luis Unit, which provided water to WWD, was authorized the same year that the California voters approved a massive state water project. Both proposals included plans for drainage facilities. The federal project planning was much further along and the state could not assure the federal government that it would be ready to participate in a joint facility when ground was broken for the San Luis Dam. When the initial bid for the project did not include any drainage features, landowners downslope from the unit sought an injunction to prevent the dam's construction. They argued that providing water without drainage would injure their lands but the suit was dismissed because the Bureau of Reclamation was showing good faith in planning the drain.

In 1964 and 1965, great controversy surrounded plans for agricultural drain water. Concerns that the lack of coordination between federal and state agencies would lead to two separate drains resulted in meetings to resolve the matter. The decision was that the state would participate in a single 467-km drainage canal from Bakersfield to Antioch Bridge. This decision raised concerns about the effects of drain water on the bay-delta system. A plan was developed that would allow construction of the initial drainage facilities, while studies were conducted on health and environmental effects of the proposal.

Thereafter the battle over the drain became budgetary. An annual rider was attached to the federal appropriations stipulating that the drain could not be completed to the Delta until certain studies had been completed. In 1967 the state decided they would not participate because of insufficient bonding capabilities and repayment commitments. Furthermore, state contractors in the southern part of the Valley had concluded that, at least for the foreseeable future, it would be more cost-effective for them to use evaporation ponds.

A plan was developed for the San Luis Drain which comprised one segment of cement-lined canal collecting and carrying drainage waters to a regulating reservoir. This reservoir, known as Kesterson Reservoir, was to be used as a wildlife habitat as well as having the principal function of drain water regulation. The San Luis Drain construction commenced in 1968 and by 1973 Kesterson Reservoir began receiving irrigation runoff waters. It was not until 1978, however, that subsurface drainage effluent was diverted to it, and by 1981 the entire flow into the reservoir was subsurface drainage effluent.

In the meantime, construction on the second segment of the canal was delayed because of the requirement for an environmental impact statement in compliance with the National Environmental Policy Act, which had become law in January 1970. Budget issues again rose. A number of facilities including the drain were pinpointed as needing reauthorization because their design and cost had significantly changed since the units' original approval 18 yr before. Concrete lining of the drainage canal and the addition of a regulating reservoir had increased the estimated cost of the drain from $7.2 to $134.9 million between 1956 and 1977.

Despite the delays and the need for draining additional lands, the condition was considered temporary as it was expected that approval and construction of the drain to the Delta would be forthcoming. That expectation was dashed by some unanticipated observations and findings. Field observations at the reservoir showed a very high incidence of mortalities and deformities among birds using the habitat. Tissue analysis revealed that Se concentrations were several times higher than found elsewhere, and Se was considered to be the primary cause of wildlife damage.

These findings revealed a problem no longer restricted to concerns about water table elevations and salt balance, and now included toxic chemicals and their effects on the environment. In March 1985 the U.S. Department of the Interior announced plans to close Kesterson Reservoir and the San Luis Drain immediately, and to terminate irrigation deliveries to 17 010 acres in Westlands because of possible criminal violations under the Migratory Bird Treaty Act. This announcement was devastating to the agricultural community that already had cropping plans and, in some cases, completed spring planting. By April an agreement had been reached between the U.S. Department of the Interior and the Westlands Water District to continue delivery of irrigation water to all of Westlands and to close Kesterson Reservoir and San Luis Drain by June 1986. By May 1986 all drains in Westlands Water District had been plugged and subsequently Kesterson Reservoir has been dried and reclamation practices completed. All that remains is a dry, 132-km long concrete canal gathering wind-blown dust and debris and serving as a monument to a 20th Century irrigation project that failed to properly accommodate problems associated with water quality. The struggle continues and the long-term consequences to agricultural productivity in the area are in question.

QUALITY OF LAND

Hillel (1991) stated that there can be no sustainable agriculture on rain-fed uplands unless erosive powers of rainstorms can be controlled and penetration of water into soil rather than runoff could be promoted. These same two factors, erosion control and water penetration, are also vital to the sustainability of irrigated agriculture. Both factors are associated with quality of land. Only soil physical properties, as characteristics of land quality, will be considered in this section. Soils laden with salt would be considered as having poor quality. Moreover, a direct linkage exists between water quality and soil physical properties, and evaluation of that linkage will be emphasized.

Reduction of erosion and enhancement of infiltration rates are complementary goals. The surface soil generally has lower bulk density and more aggregation than subsurface horizons. The dynamic biological interactions of root growth and decay, and microbial populations decomposing organics and creating byproducts that provide aggregate stability, are most prevalent in the surface layers. Wetting and drying, freezing and thawing are physical forces creating aggregation, which are also more prevalent in the upper layers. Erosion is a process by which surface soil is washed away, reducing the thickness of the A horizon, and bringing the B horizon closer to the surface.

Erosion is initiated by the overland flow of water, which acts as the transporting medium. The rate of surface water flow is inversely related to the infiltration rate. The well aggregated, low bulk density surface soil typically has higher infiltration rates than subsurface horizons. An unstable positive-feedback mechanism can be initiated. Erosion removes surface soil that exposes subsurface layers of lower infiltration rate. The lower infiltration rate results in more runoff with higher erosive capacity, which in turn removes more soil exposing layers with even lower infiltration rate. Thus, the dual goal of reducing erosion and enhancing infiltration rate as vital factors in sustainability of agriculture is apparent. Additionally, the plant nutritional aspects of surface soils surpass those of subsurface layers, so crop production is also directly impacted by these factors. For example, Carter et al. (1985) studied 14 farmer-operated fields in Idaho. Furrow irrigation erosion redistributed topsoil by eroding the upper end of the field and depositing sediment on the downslope portion of the field, causing a several-fold topsoil depth differential. Crop yield was sharply decreased in the upper end of the field where the topsoil depth was <38 cm.

The implication of water quality, specifically salinity, on agricultural productivity presented in the Water Quality section focused on direct effects of salinity on crop response. Water quality affects soil physical properties, and thus may impact agricultural production through its effect on land quality. Shainberg and Letey (1984) and Shainberg and Singer (1990) reviewed the literature on the effects of salinity and sodicity on soil physical properties. Soil infiltration rate and soil hydraulic conductivity are affected by the sodium adsorption ratio (SAR) and total electrolyte concentration of the irrigation water. The total electrolyte concentration of the irrigation water is commonly characterized by the electrical conductivity (EC) of the water. In general, increasing SAR or decreasing EC of the irrigation water tends to decrease hydraulic conductivity and infiltration rate. Saline waters can be applied to soils without destroying their soil physical conditions because the high electrolyte concentrations negate the effects of the high SAR value of the water.

Irrigation with saline waters, however, does potentially lead to poor soil physical conditions. As water of a given chemical composition becomes concentrated through evaporation, the SAR of the water increases. This effect is further enhanced if the solution becomes sufficiently concentrated so that the divalent cations precipitate. Thus, waters with high salinity also tend to be high in SAR. As water is applied to the soil, the exchangeable sodium percentage (ESP) on the soil exchange sites equilibrates with the SAR of the irrigating water. Thus, waters with high SAR lead to soils with high ESP. The high ESP is not detrimental to hydraulic conductivity as long as the percolating solution is also high in total electrolyte concentration. Almost all irrigated regions of the world, however, are subject to small amounts of rainfall during the course of the year. Since rain is essentially distilled water, the electrolyte concentration in the soil solution is reduced; and the effects of the high ESP become manifest in clay swelling and dispersion and formation of a dense, hard crust. Dispersion and crust formation are not reversed without some mechanical manipulation of the soil. Thus, the use of saline waters on soils can lead to significant destruction of the soil physical properties, the consequences of which may not be easily reversed.

Irrigation with saline waters may impact crop production more through the indirect effect on soil physical conditions than on the direct osmotic effects on plant growth. For example, Rains et al. (1987) conducted a study where irrigation waters of different salinities were applied to plots seeded to cotton. The most negative effect of the high saline waters was on the destruction of soil physical properties which reduced crop yields primarily due to a poor stand establishment. The destruction of the soil physical properties was the result of the rainfall and pre-irrigation with waters of low salinity on soils that had been irrigated with high saline waters during the crop season.

Increasing demands for water and limited supplies of fresh water provide an incentive to use saline waters for irrigation. Use of drainage water for irrigation is being proposed where other disposal options are limited. One enticing approach to drain water disposal is to sequentially apply it to more salt-tolerant crops in a manner to reduce the drainage volume and increase its concentration in each sequential use. The concept is to end up with a low volume of extremely saline water for disposal by other means.

This approach is appealing because it appears to obtain the ultimate utility of the water before disposal. The hazard of this approach is that once lured into the system a series of events may be initiated from which it will be difficult to retract. The following chain of events is probable. As water is applied, the salts become concentrated in the root zone through evapotranspiration. Even the most salt-tolerant crops have an upper level of tolerance. Additional water must be applied for leaching. Since the purpose of the operation is to dispose water through evapotranspiration, the crop must be selected to have a high ET rate. Some crops may be tolerant to salinity because they have mechanisms that restrict transpiration and therefore water uptake. Such plants would not do for the drainage disposal objective. Therefore to have high ET and yet considerable leaching to keep salinity in a tolerable range for the plant, considerable water must be applied to the land. This is acceptable because with more water application, less land is required to dispose a given quantity of drainage water. As the soil physical conditions deteriorate and the infiltration rate decreases, however, the ability to get water into the soil to provide leaching may become the limiting factor. As such, the salinity in the root zone increases and the evapotranspiration decreases. The net effect is to have reduced ET per unit land area with time. Having started out with a given amount of drainage water to be disposed, additional land must now be assigned for drainage water disposal. This process creeps along with an ever-increasing amount of land being destroyed physically to the point where very little water can infiltrate, and with little infiltration very little vegetation is sustained. The long-term effect could be a human-induced process of desertification.

SOCIETAL FACTORS

The quantity of water allocated to irrigation is a societal decision. Construction of dams, reservoirs, and canals has reduced surface outflows from geographic areas, thus increasing the quantity of water available for irrigation. Previously, attitudes have been that human demands for agricultural or urban purposes have highest priority, sometimes to the detriment of fish and wildlife. A shift in priorities where agricultural uses are being moved downward is evident in some parts of the world. For example, U.S. federal legislation passed in 1992 amended the Central Valley Project Act of 1937 in California and specifically names mitigation, protection, restoration, and enhancement of fish and wildlife as a project purpose. The Act established a $50 million annual habitat restoration fund and allocated annually 800 000 acre-feet (600 000 in a dry year) of water to the environment. In essence, the environment is now a contractor of the project. Furthermore, the legislation created the opportunity for voluntary water transfers to metropolitan and other uses.

Representative George Miller, a leader in the Act's passage, stated, "This legislation is the culmination of a decade and a half of sometimes angry effort to eliminate abuses, reduce unwarranted subsidies, modernize the operations and ameliorate the damages caused by the CVP" (McClurg, 1993). This attitude contrasts with that expressed in the *Fresno Bee* newspaper at the time the San Luis Project was being considered. "The West Side will support many thousands of families to the benefit of our state and the good of our nation . . . It will sustain homes and schools and churches and industry . . . It would represent an investment in people—it would strengthen and enhance the thing we all hold most precious—our American way of life."

Protection of fish and wildlife may impact irrigated agriculture by virtue of water quality as well as water quantity. Water salinity was previously discussed as a water quality factor limiting irrigation because of the direct effect on crop production. Some chemicals dissolved in water may not affect water use for irrigation, but can be harmful to fish and wildlife or domestic use. Agrichemicals such as fertilizers and pesticides can be transported by irrigation return flows to surface streams or aquifers. Harmful chemicals such as Se or As inherent in some soils can be mobilized by irrigation return flows.

A societal decision to protect surface or groundwaters from chemical degradation associated with irrigation return flows could significantly affect the future of irrigated agriculture. Return flows are a natural consequence of irrigation, albeit the magnitude is sensitive to management alternatives. Agricultural lands underlain by high water tables requiring drainage will be affected sooner and more stringently than other lands. Drain water discharge on the surface can be readily observed and its impact on surface waters quantified. Regulations prohibiting discharge of drainage water to streams or lakes can be enforced. The Westlands Water District in California is an example of this phenomenon, which was presented above. In this case, farmer options for drain water disposal are facing additional restrictions. The farmer could conceivably sacrifice land to establish evaporation ponds, but toxic elements such as Se can become concentrated as the water evaporates to the level that the pond becomes classified as a hazardous waste. Disposal of hazardous waste is rigidly regulated and requires very expensive mitigation practices.

Subsurface flows that discharge into streams or aquifers are difficult to quantify or regulate. Enforcement of any regulation that might be adopted is likewise difficult. Irrigation on these lands is less likely to be immediately impacted than land with high water tables. But, nevertheless, agriculture is vulnerable to approaches that might be taken to protect water quality. Frustration in developing suitable mitigation practices could lead to the drastic decision to prohibit irrigation to eliminate the source of water degradation.

The main point is that sustainability of irrigated agriculture is not restricted to physical-biological constraints, but is highly dependent on societal attitudes. These attitudes will vary with locale and time and prediction of the long-term consequences would require prophetic vision.

ECONOMIC FACTORS

Many of the physical-biological constraints on irrigated agriculture can be mitigated with sufficient financial investment. Water quality is a major factor in irrigated agriculture. Water quality can be changed through treatment so it becomes a matter of economics. For example, some oil-rich and fresh water-poor Arab countries have expended oil to convert salt water to fresh water through desalinization. Capital gains from the sale of oil and energy from the oil are used to construct and operate the desalinization plants. One point of view is that water quality is not a physical or biological constraint, but rather an economic constraint. At present, investment in a desalinization plant for irrigation is not economically justified because the returns do not offset the costs.

Desalinization was considered as one option of managing saline drainage waters in the western San Joaquin Valley of California. The plan included using the concentrated brine solution from the desalinization plants to develop solar power ponds (Letey et al., 1986). Desalinization of drainage water through reverse osmosis was found to have more technological difficulties than desalinization of seawater. The relatively high concentration of Ca and SO_4 in the drainage water caused $CaSO_4$ to precipitate as the solution was concentrated during the reverse osmosis process and the precipitate clogged the membrane. Thus a water softening pretreatment of the water was necessary prior to reverse osmosis. Due to technological and economic constraints, this option has not been adopted for the management of drainage water in the Valley.

Less dramatic approaches than building desalinization plants are possible. The major negative impact of irrigated agriculture is the increase in salinity and the development of water tables near the surface. The rate of development of both of these factors can be moderated through irrigation management. Irrigation management includes the timing, amount, and method of irrigation water application. The primary goal of irrigation management is to achieve high crop yields. Irrigation management to maximize the potential for sustainability of irrigated agriculture would consider both crop yield and the amount of water percolating below the root zone. The water percolating below the root zone is a major contributor to the buildup of water tables and development of soil salinity. Some deep percolation is an inevitable consequence of irrigation. The magnitude of deep percolation, however, is amenable to control through management.

Crop yield and deep percolation are functions of the amount and uniformity of water application. The degree to which these two entities can be controlled is partially dependent upon the irrigation system. Irrigation systems can be broadly classified as being either gravity-flow or pressurized. Gravity-flow (surface) systems are characterized by water flow in channels across the field. A channel may be a furrow between crop rows, a strip of land bordered by low dikes, or an entire field. The amount and uniformity of water infiltration for gravity-flow systems are largely functions of soil characteristics, although the irrigator has control of when to release and cut off water supply to the field. Pressurized systems deliver water under pressure through pipes and release it from sprinkler nozzles, small orifices, or tubes. Well-designed pressurized systems transfer control of infiltration quantity and uniformity from the soil characteristics to the design and maintenance of the delivery system. Pressurized irrigation systems give the irrigator precise control over the amount of applied water because it is delivered through pipe with a valve that can be controlled at will. The uniformity of irrigation is controlled by the design and maintenance of the delivery system. Sprinkler irrigation uniformity is significantly affected by wind currents which, of course, are not under the irrigator's control.

Investment costs for pressurized systems are considerably higher than for gravity-flow systems. Thus, from an economic point of view, the increased costs must be offset by comparable or greater benefits. An economic analysis of irrigation systems for cotton production in California was reported by Letey et al. (1990). With no cost or restrictions to the farmer on drainage water disposal, the gravity-flow systems were more profitable than the pressurized systems. If costs for drainage water disposal were imposed on the farmer, however, a point would be reached where a switch from a gravity-flow to a pressurized irrigation system would be economically justified to the farmer.

Soil erosion is another factor that can impact the long-range productivity of irrigated agricultural lands. Water flow across the land surface acts as the transporting medium for erosion. Gravity-flow irrigation systems require water flow over the land surface and contribute to erosion. In contrast, pressurized irrigation systems designed to apply water at rates equal to or less than the infiltration rate prevent surface runoff, thus mitigating erosion. Investment in irrigation systems that prevent erosion can contribute long-term benefits to agriculture. A dramatic example of the feasibility of switching to pressurized irrigation systems can be observed by driving through southern California where the rather steep hills have been completely covered by avocado (*Persea americana* Miller) trees. Crop production on such topography would be impossible without the advent of pressurized low-application irrigation systems that prevent runoff.

Sustainable agriculture to some is interpreted to mean low-input agriculture or a return to agricultural practices prevalent decades ago where crop rotation and nutrient recycling were practiced. From an irrigation point of view, the higher technology irrigation systems have a higher potential for maintaining agricultural production by decreasing the impacts of erosion, water build-up, and the development of salinity. The short-term benefits to the farmer, however, may not be perceived to justify the investment. The long-term costs and environmental degradation are usually not factored into farmer decisions. Furthermore, some of the environmental degradation costs have not been imposed directly on the farmer, thus reducing the incentive for a farmer to invest in irrigation systems to reduce the degradation.

One definition of agricultural sustainability is "staying in business." Thus, the future of irrigated agriculture is very much related to economic factors. One simple economic principle is related to supply and demand. As supply becomes limited, price goes up. There are a number of obvious examples where this principle applies. The Organization of Petroleum Exporting Counties (OPEC) can partially control oil prices by decisions on the amount of oil placed on the market. Value of diamonds is maintained by limiting the supply. The question is whether this principle is valid when it is applied to food production. People can survive without oil or diamonds, but they cannot survive without food. Starving people are starving because they have limited resources to purchase food. In a completely competitive situation, as populations increase and food demands increase, the price of food would go up. As food prices rise, the farmers have more resources to invest in production activities such as upgrading irrigation systems. This would appear to be a positive factor in the sustainability of irrigated agriculture. As prices go up, however, poor people have even less ability to purchase food and their plight is made worse. A portion of the population could die from starvation. These deaths result in less demand for food, so ultimately a theoretical equilibrium point is reached between food supply, food prices, and the surviving population size.

For humanitarian and other reasons, a completely competitive market system is not likely. Ample examples exist of cooperative worldwide efforts to provide food for starving populations in some parts of the world. Thus, the question remains as to what extent increased food demands caused by population increases will lead to higher farm income such that investments can be made for water treatment or technological advances in irrigation. The answer to this question is an essential factor in the sustainability of irrigated agriculture, and I am not competent to evaluate it.

Survival is a strong biological instinct that can drive the future of irrigated agriculture in various directions. Allocation of water supplies between irrigated agriculture and fish and wildlife might shift towards agriculture as increasing food demands lead to hunger in the population. Such a shift, if it occurs, would benefit irrigated agriculture, but society would bear the costs associated with diminished fish and wildlife resources. Severe food shortage could lead to practices that are detrimental to sustainable irrigation. Just as starving animals not only eat leaves of plants, but also chew on the branches and roots, thus diminishing the capacity of the plant to produce; starving humans may diminish long-term productivity in an attempt for short-term returns. Such practices include using waters with chemical composition which, with time, could destroy physical properties, or exploiting stored water resources at a more rapid rate than they are being restored.

In conclusion, irrigated agriculture is sustainable if appropriate long-range planning and investments are imposed. If history is a reflection of the future, the outlook is rather bleak. Groundwater resources are being exploited at rates considerably higher than the sustainable recharge. Irrigation projects are initiated without adequate planning and development of drainage systems. Salinization of lands is progressing at an unabated rate. Yet there is reason for optimism. The physical-biological principles are reasonably well understood. There are management options that promote the sustainability of irrigated agriculture. What is needed is the will of the public to adopt and invest in those practices which will lead to sustainability of agriculture.

A most depressing message was given by a professor whom I invited to give a guest lecture to my class when he stated that there will always be enough food to feed the peoples of the world. The "excess" population will die until the statement is true. Most would prefer a mechanism to bring population and food supply in balance by means other than premature death. Developing a sustainable irrigated agricultural production system is a key component of this goal. Success can come from the societal will to apply established physical-biological scientific principles to plan and invest for a sustainable future.

REFERENCES

Carter, B.L., R.D. Berg, and B.J. Sanders. 1985. The effect of furrow irrigation erosion on crop productivity. Soil Sci. Soc. Am. J. 49:207–210.

Dorsey, C.W. 1906. Alkalized soils of the United States. USDA, Bureau of Soils Bull. 35. U.S. Gov. Print. Office, Washington, DC.

Hilgard, E.W. 1886. University of California, College of Agriculture, Report to the President of the University. California State Printing Office, Sacramento.

Hilgard, E.W. 1889. The rise of the alkali in the San Joaquin Valley. Univ. of California Exp. Stn. Bull. 83. Univ. of California, Berkeley.

Hillel, D.J. 1991. Out of the earth. The Free Press, New York.

Letey, J. 1993. Relationship between salinity and efficient water use. Irrig. Sci. 14:75–84.

Letey, J., A. Dinar, C. Woodring, and J. Oster. 1990. An economic analysis of irrigation systems. Irrig. Sci. 11:37–43.

Letey, J., C. Roberts, M. Penberth, and C. Vasek. 1986. An agricultural dilemma: Drainage water and toxics disposal in the San Joaquin Valley. Univ. of California Agric. Exp. Stn. Spec. Publ. 3319. Univ. of California, Oakland.

McClurg, S. 1993. Changes in the Central Valley Project. Jan.–Feb. p. 4. Western Water. Water Education Foundation, Sacramento, CA.

Rains, D.W., S. Goyal, R. Weyrack, and A. Lauchli. 1987. Saline drainage water reuse in a cotton rotation system. Calif. Agric. 41(5):24.

Shainberg, I., and J. Letey. 1984. Response of soils to sodic and saline conditions. Hilgardia 52(2):1–57.

Shainberg, I., and M.J. Singer. 1990. Soil response to saline and sodic conditions. p. 91. *In* K.K. Tanji (ed.). Agricultural salinity assessment and management. ASCE Manuals and Rep. on Eng. Practice no. 71. ASCE, New York.

5 Global Overview of Soil Erosion

Rattan Lal

Ohio State University
Columbus, Ohio

The world's cropland area has not increased since 1980 (Brown, 1991a) and is constant at $\approx 1.4 \times 10^9$ ha for a total population of 5.4×10^9 and agricultural population of 2.4×10^9 (Food and Agriculture Organization, 1991). In fact, per capita arable land area is rapidly decreasing (Table 5–1). Most populous countries of the world are losing their precious cropland to soil erosion and other degradative processes. Soil erosion is a major factor in land:population ratio in these countries. The Loess Plateau of China is losing soil at an alarming rate exceeding 100 Mg ha^{-1} yr^{-1} (Dazhong, 1993). Shiwalik hills and lower Himalyan regions of India are losing top soil at a rate in excess of 80 Mg ha^{-1} yr^{-1} (Singh et al., 1992).

There are several ways to address the issues relevant to a global soil erosion. One is to survey the magnitude of soil erosion by different processes. Another is to discuss its importance in terms of geologic, economic, and environmental impacts. And yet another approach is to critically review the available data, discuss their uses and limitations, and identify knowledge gaps for future research and development. It is often difficult to address the soil erosion issue objectively because of (i) different goals for collating the statistics on soil erosion, (ii) different terminology used to denote similar processes, and (iii) lack of standardized methods used in data procurement and analyses. Agriculturists are often interested in short-term rate of erosion in relation to a farm field's cropping or farming system. In contrast, geologists and hydrologists are concerned with long-term rate of erosion, across a range of slopes, and over large areas involving major river basins. These data from farms on hill

Table 5–1. Per capita arable land area of the world (calculated from Food and Agriculture Organization, 1991).

Year	World	Asia	China
	ha		
1975	0.33	0.18	0.100
1980	0.30	0.16	0.093
1981	0.28	0.15	0.088
1990	0.25	0.14	0.082

Soil and Water Science: Key to Understanding Our Global Environment, SSSA Special Publication 41.

slopes and those from river basins are often not comparable and lead to conflicting interpretations.

Soil erosion, displacement of soil material by agents of erosion (e.g., water, wind, and gravity), is a natural geological process of landscape development. Occurring at the natural geological rate, formations of some of the most productive and fertile soils of the world are credited to soil erosion and its positive effects on landscape evolution. Human-induced perturbations, however, accelerate the soil erosion rate. The accelerated erosion can have deleterious effects, especially when the rate exceeds the natural geological rate over an extended period of time. Therefore, an objective assessment of soil erosion can only be made by understanding the accelerated rates of erosion for different physiographic and climatic environments, and land uses and farming systems.

Global assessment of soil erosion also involves a thorough understanding of the processes at different scales, e.g., aggregate level, hillside, landscape, watershed, and river basin. Processes of soil erosion and sediment transport differ at each scale. Ideally, the processes involved, the magnitude of soil displaced, and interrelationship between soil displaced and that transported out of the field should be known for each scale. Different terms used to describe these processes are described below:

1. Erosion refers to total erosion caused by raindrop impact and thin overland flow. It is also called interrill-rill erosion and involves erosion due to processes at levels of aggregate and small plots along hillside. Gross erosion is usually measured using field plots of 10 to 100 m^2 size and may be of the order of several hundreds megagrams per hectare per year.
2. Soil Loss is the net amount of soil moved off a particular field or area. It is equal to total erosion of soil dislodged by raindrop impact and overland flow minus deposition or sedimentation in micro-depressions within the small field.
3. Sediment Yield refers to the total sediment flow at a specific point in the landscape. Sediment yield is the sum of soil erosion from each slope segment within the landscape or watershed minus the deposition. The deposition usually occurs in depressions, at the toes of slopes, and in channel terraces. Total sediment yield involves both suspended and dissolved (solution) material. Sediment yield is measured on watersheds of one to several hundreds of hectares.

Considering the magnitude of these three terms, gross or total erosion > soil loss > sediment yield. The ratio of magnitude of sediment outflow from larger to the smaller scale is called *delivery ratio.*

4. Denudation Rate refers to the rate of transport of suspended solids and dissolved solutes from a watershed outlet expressed per unit area of the watershed. It is expressed either on a depth (mm yr^{-1}) or a volume (m^3 km^2 yr^{-1}) basis. Denudation rate is usually calculated for large river basins of several thousands of kilometers per square meter, e.g., Mississippi River Basin with a denudation rate of 50 to 60 mm per 1000 yr (Schumm & Harvey, 1982).

These four terms apply mostly to erosion by water. In contrast to water erosion, it is even more difficult to quantify erosion by wind at different scales.

GLOBAL WATER BALANCE

The global magnitude of soil erosion is usually estimated from the amount of sediment transport to the oceans. All other factors remaining the same, sediment transport to the oceans depends on runoff rate and amount. In fact, accurate assessment of the sediment transport to the ocean depends on reliable estimates of global runoff. USSR National Committee for the International Hydrological Decade (USSR, 1974) estimated that global runoff from all land area is about at 47 x 10^3 km^3 or 9.2 cm. The exact magnitude, however, is influenced by human activities, e.g., storage in reservoirs (≈5 x 10^3 km^3), irrigation (≈4 x 10^3 km^3 for 300 x 10^6 ha of cropland), and groundwater recharge (United Nations Scientific and Cultural Organization, 1978; Walling, 1987). Management of global runoff can have a drastic effect on hydrologic cycle and sediment transport to the oceans.

MAGNITUDE OF SOIL EROSION

Soil Erosion and Soil Loss

Field data on soil erosion and soil loss are usually obtained for site-specific experiments on field runoff plots conducted on a range of soil types or phases of series, terrain characteristics, vegetation and soil cover, and rainfall regimes. A first approximation of the erosion rate, whatever it is worth, can be made by developing a data bank of erosion for principal soils in major ecoregions of the world. A well planned and coordinated effort is needed to achieve this because in the tropics alone there are 11 soil orders, 45 suborders, 200 great groups, 1250 subgroups, 1 x 10^6 families, 5 x 10^6 series, and 10 x 10^6 phases of soil (Eswaran et al., 1992). Even if these data were available, however, it would be difficult to translate these data into accumulative erosion rates over the land area of the earth.

In the absence of precise data on soil erosion or soil loss, it is common to present the land area affected by different types of soil erosion. An example of this type of information regarding the global extent of soil erosion is shown in Table 5–2 (International Soil Reference and Information Center/United Nations Environment Program, 1990). Out of the total degraded land, area affected by water erosion covers 46% of Africa, 74% of Central America, 63% of North America, 51% of South America, 58% of Asia, 52% of Europe, 81% of Oceania, and 56% of the world. Areas affected by water and wind erosion, however, cannot be added together because some regions are affected by both water and wind erosion, e.g., arid and semiarid regions of the world. A major limitation of this approach lies in the lack of information about the severity of soil erosion.

Another variant of this qualitative approach of assessing the magnitude of soil erosion is that adopted by United Nations Environment Program in assessment of desertification (Dregne, 1986; United Nations Environment Program, 1990). Severity of erosion is assessed according to different categories, e.g., slight, moderate and severe, and land area is estimated under each category.

Table 5–2. Water and wind erosion as percent of the total area affected by land degradative processes (World Resources Institute, 1992–1993).

Region	Water erosion	Wind erosion
	%	%
World	56	28
Africa	46	38
America		
Central	74	7
North	63	36
South	51	17
Asia	58	30
Europe	52	19
Oceania	81	16

Table 5–3. Soil loss in the loess plateau region of China (Jing, 1986; Dazhong, 1993).

Erosion level	Area	Soil erosion rate
	10^6 ha	Mg ha^{-1}
I	7.2	>100
II	10.8	50–100
III	5.7	20–50
IV	19.3	<20

Table 5–4. Major soil erosion regions in China (Dazhong, 1993).

Region	Total area	Eroded area	Total soil erosion
	10^6 ha	10^6 ha	10^6 Mg yr^{-1}
Loess Plateau	53	43.0	2200
Southern Region	160	69.0	2500
Northern Region	28	23.0	500
Northeastern Region	13	2.5	150

Because different categories of soil erosion are rather subjective, these estimates are also qualitative and useful perhaps for overall planning and for creating awareness about the extent of the problem. Again, the severe erosion category is subjective and qualitative. The adverse impact of severe erosion depends on antecedent conditions, land use, climate, and management. The data in Tables 5–3 and 5–4 from the Loess plateau and other regions of China are examples of semiquantitative data because different categories of erosion have been assigned a quantitative range and the corresponding area is also quantified. Similar data are also available for India with quantitative information on the magnitude of erosion and the corresponding area affected by it (Table 5–5). These data can be improved, however, if category of erosion was quantified in relation to its agronomic or economic impact.

Sediment Yield

The data on sediment transport for major rivers and their tributaries is used to compute sediment yield (Walling, 1987). Several researchers have collated the

Table 5–5. Area under different classes of soil erosion in India (Singh et al., 1992).

Category	Erosion rate	Area
	Mg ha^{-1} yr^{-1}	10^3 km^2
Slight	<5	801.4
Moderate	5–10	1405.6
High	10–20	805.0
Very high	20–40	160.1
Severe	40–80	83.3
Very severe	>80	31.9

Table 5–6. Sediment yield and denudation rates for some regions of the world.

Region	Sediment yield†	Denudation rate
	Mg km^{-2} yr^{-1}	mm yr^{-1}
Amazon Basin	50–200	0.03–0.13
Andean region	500–1000	0.33–0.67
Central America	500–1000	0.33–0.67
East African highlands	200–500	0.13–0.33
Eastern Australia & New Zealand	200–500	0.13–0.33
Europe	20–100	0.013–0.067
Himalyan-Tibetan ecosystem	1000–2000	0.67–1.33
Loess plateau, China	4000–5000	2.67–3.33
Mediterranean Basin	500–1000	0.33–0.67
Southeast Asia	500–1000	0.33–0.67
U.S. Corn Belt	50–500	0.03–0.33
West Africa	50–500	0.03–0.33
Western Australia	5–50	0.003–0.03
Western USA	500–1000	0.33–0.67

†L'Vovich et al., 1990; Bulk density of sediment = 1.5 Mg m^{-3}; 1 mm of soil = 1500 Mg km^{-2}.

available information on sediment yield for major rivers of the world (Douglas, 1973; Fournier, 1960; Holeman, 1968; Meybeck, 1976; Milliman & Meade, 1983). These results of sediment yield are often mapped as denudation rates for the region. Such global maps have been compiled by Jansson (1988) and L'Vovich et al. (1990). Comparison of the data in Tables 5–6 and 5–7 from two such global maps indicates the problem because the denudation rates for the same region are different. Similar magnitude of discrepancies and anomalies exist in other data available in the literature.

Calculations of the denudation rates from sediment yield are based on the information on bulk density of the sediment. This information is often not available and is commonly assumed. L'Vovich et al. (1990) assumed the bulk density to be 2.0 Mg m^{-3}. Bulk density of 1.5 Mg m^{-3} was used in the present calculations of denudation rates shown in Tables 5–5 through 5–8. Assumption of an average bulk density of sediment originating from large watersheds with a wide range of soils can lead to gross errors and erroneous interpretations.

Regional and Global Estimates of Sediment Transport

The data of sediment transport for river basins is used to compute regional and global estimates of dissolved and suspended sediment flow. An example of

Table 5–7. Sediment yield and denudation rates for some regions of the world (Jansson, 1988).

Region	Sediment yield	Denudation rate†
	Mg km^{-2} yr^{-1}	mm yr^{-1}
Brahma putra Basin, Southeastern China	500–1000	0.33–0.67
Canada	50–100	0.033–0.067
Central Africa	50–100	0.033–0.067
Central South America	<50	<0.033
Eurasia	<50	<0.033
Europe	50–100	0.033–0.067
Himalyan-Tibetan ecosystem, North-Central India	>1000	>0.67
Indus Valley	100–500	0.067–0.33
Loess plateau, China	>1000	>0.67
Midwestern USA	100–500	0.067–0.33
Southeastern Australia	50–100	0.033–0.067
The Amazon Basin	100–500	0.067–0.33
The Maghreb, and Andean region	>1000	>0.67
The Sahel	<50	<0.033
Western Canada	100–500	0.067–0.33

†Bulk density = 1.5 Mg m^{-3}; 1 mm of soil = 1500 Mg.

Table 5–8. Sediment yield and denudation rates for different regions.

Region	Mean sediment yield	Total sediment flow	Denudation rate†
	Mg km^{-2} yr^{-1}	10^6 Mg yr^{-1}	mm yr^{-1}
Africa‡	35	530	0.023
Asia	229	6433	0.153
Europe	50	230	0.032
North & Central America	84	1462	0.055
Oceania & Pacific Islands	589	3062	0.390
South America	100	1788	0.067
World§	118	15434	0.079

† Bulk density = 1.5 Mg m^{-3}
‡ Excluding Sahara
§ Milliman & Meade (1983); Walling (1987)

this approach for different regions of the world is shown by the data in Tables 5–8 and 5–9. The mean sediment yield is in the order of Oceania and Pacific islands > Asia > South America > North and Central America > Europe > Africa. Total sediment flow, which depends on the land area, is the highest for Asia. These results are, however, different than those reported in Table 5–9 by L'Vovich et al. (1990). The major discrepancy in the data lie in sediment flow and denudation rate for Oceania and Pacific Island region.

Because of a wide variation in methods of sampling, calculations of sediment flow, and other techniques used, there is a wide range of estimates of total transport of suspended sediments to the world's oceans. The data in Table 5–10 show that estimates of mean annual load range from 8.3 x 10^9 to 58.1 x 10^9 Mg yr^{-1} with corresponding global denudation rate from 0.04 to 0.30 mm.

Table 5–9. Contemporary rates of sediment yield and denudation rate (L'Vovich et al., 1990).

Region	Sediment yield	Total sediment flow	Denudation rate†
	Mg km^{-2}	10^6 Mg	mm
Africa	57	1720	0.028
Asia	186	7620	0.093
Australia & Oceania	48	360	0.024
Europe	73	720	0.036
North America	98	2080	0.049
South America	137	2430	0.068
World	118	14930	0.059

†Assuming bulk density of 2 Mg m^{-3}

Table 5–10. Some estimates of suspended sediment transport to the oceans.

Source	Mean annual load	Denudation rate†
	10^9 Mg	mm
Fournier (1960)	58.1	0.30
Gilluly (1955)	31.7	0.16
Goldberg (1976)	18.0	0.09
Holeman (1968)	18.3	0.09
Jensen & Painter (1974)	26.7	0.14
Keunen (1950)	32.5	0.17
L'Vovich et al. (1990)	15.0	0.075
Lopatin (1952)	12.7	0.06
MacKenzie & Garrels (1966)	8.3	0.04
Milliman & Meade (1983)	13.5	0.07
Pechinov (1959)	24.2	0.12
Schumm (1963)	20.5	0.10
Walling & Webb (1983)	15.0	0.075
Walling (1987)	17.4	0.088

†Assuming bulk density of 1.5 Mg m^{-3}; Land area = 13.1 x 10^9 ha; 1 mm loss = 196.5 x10^9 Mg.

IMPACT OF SOIL EROSION

Accelerated rate of erosion due to human-induced perturbations has major adverse effects on soil productivity on-site, and damage to waterways and civil structures off-site. In addition, there are drastic adverse effects on water and air qualities with severe global environmental consequences.

Rate of New Soil Formation

Impact of soil erosion is adverse and soil is a nonrenewable resource if rate of erosion exceeds the rate of soil formation. The rate of new soil formation is difficult to determine, not known for most geological formations and ecoregions, and varies considerably with the parent material, vegetation type, and climatic environment. It is commonly believed that the rate of soil formation is ≈2.5 cm per 1000 yr (Chamberlain, 1909; Bennett, 1939). The rate of new soil formation is generally higher for soils of volcanic origin than those derived from residual parent rock (Lal, 1994b). Friend (1992) estimated that worldwide the rate of new soil formation is ≈25 mm per 150 yr or 0.17 mm yr^{-1}.

Agronomic Productivity

Impact of soil erosion on productivity depends on a multitude of interacting factors (Fig. 5–1), and it is difficult to generalize its global effect. In some soils, erosion may have either none or a slightly positive effect on productivity. In others, adverse effects on productivity depend on management. Productivity of such eroded soils can be enhanced by improved management. In marginal soils, where accelerated erosion has reduced topsoil depth to below the critical level, management may have no impact in enhancing productivity because the soil has been irreversibly degraded.

Few attempts have been made to assess the impact of erosion worldwide on productivity. Dregne (1990) estimated that productivity of some parts of Africa has declined by as much as 50% due to soil erosion and desertification. Productivity losses due to erosion in Africa are serious in the Maghreb, Nigeria, and Ghana in West Africa, and in parts of southern Africa and east African Highlands. Dregne (1992) also observed serious productivity losses in Asia including parts of India, China, Iran, Israel, Jordan, Lebanon, Nepal, and Pakistan. Productivity loss in several parts of Asia is estimated to be as much as 20% due to erosion-induced soil degradation. Lal (1995) estimated that mean productivity loss due to past erosion in Africa was 9% for sub-Saharan Africa with a range of 2 to 40%. If accelerated erosion continues unabated, mean productivity decline in sub-Saharan Africa may be 14.5%. Annual reduction and total production in sub-Saharan Africa in 1989 was estimated at 3.6 x 10^6 Mg for cereals, 6.6 x 10^6 Mg for roots and tubers, and 0.4 x 10^6 Mg for pulses. In severely affected ecoregions, erosion-induced productivity decline may be a principal factor of perpetual food deficit in sub-Sahran Africa (Table 5–11; Lal, 1995).

Similar to assessing total soil erosion on a global scale, it is also difficult to assess productivity decline due to global soil erosion. In fact, there are no standardized methods to assess the erosion-caused loss in productivity at the global

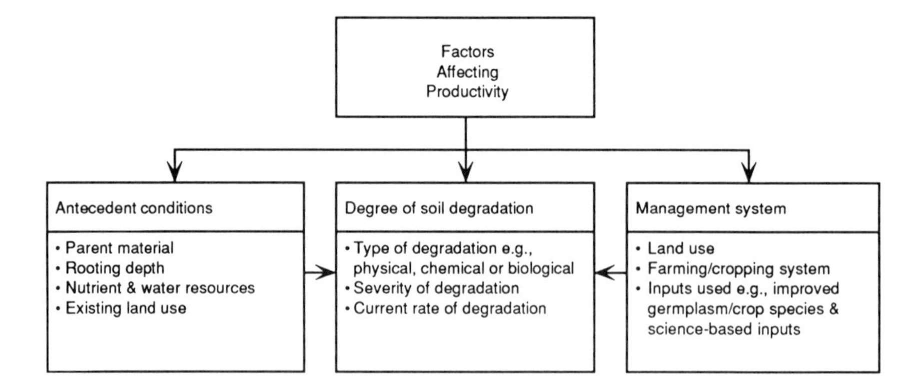

Fig. 5–1. Effect of erosion on soil productivity.

Table 5–11. Erosion effects on food production in sub-Sahara Africa (Lal, 1995).

	Actual production in 1990		Production if there were no erosion	
Commodity	Yield	Total production	Yield	Total production
	kg ha^{-1}	10^6 Mg	kg ha^{-1}	10^6 Mg
Cereals	1140	55.0	1215	58.7
Pulses	514	5.5	548	5.8
Roots and tubers	6460	98.1	6886	104.5

scale. Field experiments to assess productivity loss are done on soils whose management history and the magnitude of past erosion are known. The data from such experiments can be extrapolated to soils with similar properties and management history around the world. Loss in potential productivity can also be estimated from erosion-induced changes in soil properties. Several models are available to assess productivity effects, e.g., productivity index (PI) or erosion productivity impact calculator (EPIC). Application of such models depends on the availability of baseline data and adaptation to soil-specific conditions. If the magnitude of soil loss and its composition are known, productivity loss can also be assessed from the loss of equivalent amount of nutrients and water to produce the yield level equivalent to those from a noneroded soil. Most of these methods produce general indications of possible trends rather than actual loss of productivity due to erosion.

Water Quality

Erosion impacts water quality through transport of sediments and sediment-borne pollutants. Soil erosion increases eutrophication of water and accentuates environmental pollution. The data in Table 5–12 show that global transport of dissolved load to the oceans is about 4 x 10^9 Mg yr^{-1}. In addition, sediment laden chemicals may account for an equivalent if not more quantity of chemicals transported in the world's rivers. Impact on water quality is a major off-site effect of accelerated erosion. Magnitude of the economic and environmental impacts of adverse water quality due to accelerated erosion, however, are difficult to assess.

The Greenhouse Effect

Soil erosion is a major factor affecting global C balance. The amount of C displaced by soil erosion, however, is not known. World soils contain ≈1500

Table 5–12. Dissolved load transported from different continents to world's oceans (Walling, 1987)

Continent	Dissolved load
	10^6 mg yr^{-1}
Africa	201
Asia	1592
Europe	425
North and Central America	758
Oceania/Pacific Islands	293
South America	603
Total	3872

x 10^{15} g C in the top 1-m depth. The global sediment transport to the oceans of 19 x 10^{15} g yr^{-1} (Walling, 1987) is equivalent to 190 x 10^{15} g yr^{-1} of soil displaced from terrestrial ecosystems assuming the mean delivery ratio of 10%. With a mean C content of 3%, total C displaced in soil from the terrestrial ecosystems is 5.7 x 10^{15} g yr^{-1}. Assuming that 20% of the C displaced is decomposed and emitted into the atmosphere, C flux into the atmosphere from soil physically displaced by erosion is about 1.14 x 10^{15} g yr^{-1}. This amount is about one-third of the C emitted by burning fossil fuel.

Similar calculations can also be made for soils of the tropics. Lal (1994a) estimated that total C displaced by soil erosion in the tropics is 1.6 x 10^{15} g yr^{-1}. Assuming 20% of the displaced C is decomposed, the C flux due to soil erosion in the tropics is $\approx$0.3 x 10^{15} g yr^{-1}. These figures are at best estimates, and merely indicate the importance of soil erosion on global C balance and possible impact on C emission into the atmosphere.

LIMITATIONS OF THE DATA BASE

It is difficult to objectively evaluate the magnitude or severity of the soil erosion problem from the available statistics at global scale. Estimates of soil erosion and desertification are alarming. There, however, are several problems with these statistics, including the following:

Data Credibility and Reliability

One of the major problems with soil erosion is the data reliability. Very often, data on soil erosion are available even for developing countries and remotely accessible regions. The major problem is in the quality, accuracy and reliability of the available data. The data are often collected by wrong or uncalibrated equipment, incorrectly analyzed, or simply guessed. Global extrapolations on the basis of data collected by diverse and nonstandardized methods can lead to gross errors and erroneous and misleading interpretations. Is unreliable information better than no information? Estimates often differ by a factor of 3 to 5. Such statistics, however, can play an important role in creating awareness about the problem. If the statistics of high estimates are correct, the challenge they present to the human race is one of the greatest because soil resources are finite and nonrenewable and are rapidly being depleted. If the statistics of low estimates are correct, it is still a matter of urgency for policy makers to do something about it. If none of the estimates are even approximately correct, however, we have a serious credibility problem.

Methodology

One of the problems with the statistics is the methodology. There is no standardized methodology to obtain reliable data even on plot scale or hillside. Delivery ratio is the biggest unknown at the scale of landscape or watersheds. Methods used are decided by the financial support available. Data are often collected with improper or improperly calibrated equipment. Problems with methods are outlined in Table 5–13. There are similar problems with methodology to

Table 5–13. Minimum data set needed at different scales.

Scale	Data needs
Plot	Antecedent soil moisture, basic soil data, climatic erosivity, soil splash, soil surface conditions, and slope characteristics
Agricultural Watershed	Runoff rate and amount, sediment concentration, soil and climatic data, land use, farming system, drainage density, parent rock, geology, and terrain characteristics
River Basin	Delivery ratio, geology, predominant land uses, climate, water balance, dissolved, and suspended load
Global	Runoff rate and sediment load for major rivers, erodibility of world soil orders, climatic erosivity of principal ecoregions, and population density

assess the erosion-productivity relationship. Information on soil erosion–crop productivity can be improved by defining critical limits of soil properties in relation to different categories of soil erosion. Critical limits of soil properties imply threshold values beyond which soils' life support processes are severely jeopardized. In relation to soil erosion, important properties whose limits need to be defined are rooting depth, soil organic C, soil structure, intensity and capacity factors for nutrients and water, etc. While modeling can be a useful tool, developing accurate models is also constrained by quality and reliability of the available data.

Scaling

There is a major problem with scaling. Soil erosion and hydrological processes are scale-dependent. How can the data obtained at plot scale be extrapolated to landscape, watershed, or river basin scale? Similarly, how can the data obtained at the scale of river basin be interpolated to the field scale? What are the processes involved at different scales, and how can observations made across a range of heterogeneous scales be compared? Despite considerable interest in soil erosion research, scaling rules have not been developed and limits to extrapolation are difficult to apply. High priority should be given to developing methodologies for extrapolating erosion and hydrological processes from one scale to another.

RESEARCH NEEDS

There exists a vast amount of literature on soil erosion and its impact on productivity, agricultural sustainability, and environmental quality. Some of the available information is emotional rhetoric, some of it is confusing, and other is erroneous and misleading. An important priority is to improve credibility and reliability of the data available in the literature. Towards that goal, researchable priorities have been identified.

Standardized Methodology

There is an urgent need to improve the global data base, its accuracy and reliability. To do this is to standardize the methodology and adopt strict quality control measures. The statistics to be published must be judged by the methods used.

Socioeconomic Aspects

In addition to biophysical factors affecting erosion, it is important to relate soil erosion to socioeconomic factors, e.g., demographic pressure, educational level, and standard of living. Dazhong (1993) related increase in soil loss over time in the Loess plateau of China to the increase in population of the region. He observed a very close relationship between the total population of the region and the soil loss. There is a need to develop similar relationships for other principal river basins albeit at different levels of science-based inputs.

Soil Erosion and Crop Productivity

Conceptual and empirical models are needed to relate soil loss to crop yield at different levels of management. Unless reliable relationships between erosion and productivity are developed, it is difficult to plan development strategies. It is important to establish numerical limits of soil loss tolerance for major world soils on the basis of erosion-productivity relationship for a range of input and management systems.

Restoration of Eroded Lands

Bringing degraded lands into production is a high priority, and methods need to be developed to restore eroded lands. Land evaluation criteria should be developed to indicate the time when land should be taken out of production and put under a restorative and ameliorative phase.

REFERENCES

Bennett, H.H. 1939. Soil conservation. McGraw Hill Book Co., New York.

Brown, L.R. 1991. The global competition for land. J. Soil Water Conserv. 46:394–397.

Chamberlain, T.C. 1909. Soil wastage. *In* Proc. of Conf. of Governors in the White House, Washington, DC. 1908. U.S. Congress 60th. 2nd Session. House Document 1425. U.S. Congress, Washington, DC.

Dazhong, W. 1993. Soil erosion and conservation in China. p. 63–85. *In* D. Pimental (ed.) World Soil Erosion and Conservation. Cambridge Univ. Press, England.

Douglas, I. 1973. Rates of denudation in selected small catchments in eastern Australia. Univ. of Hull, Occasional Pap. in Geography 21. Univ. of Hull, England.

Dregne, H.E. 1986. Extension and distribution of desertification process. p. 10–16. *In* Reclamation of arid territories and combating desertification: A comprehensive approach. Center for International Projects, Moscow.

Dregne, H.E. 1990. Erosion and soil productivity in Africa. J. Soil Water Conserv. 45:431–436.

Dregne, H.E. 1992. Erosion and soil productivity in Asia. J. Soil Water Conserv. 47:8–13.

Eswaran, H., J. Kimble, and T. Cook. 1992. Soil diversity in the tropics: Implications for agricultural development. p. 1–16. *In* R. Lal and P.A. Sanchez (ed.) Myths and science of soils of the tropics. SSSA Spec. Publ. 29. SSSA and ASA, Madison, WI.

Food and Agriculture Organization, 1991. Production yearbook. Vol. 45. FAO, Rome.

Fournier, F. 1960. Climat et erosion. Presses Universitaires de France, Paris.

Friend, J.A. 1992. Achieving soil sustainability. J. Soil Water Conserv. 47:156–157.

Gilluly, J. 1955. Geologic contrasts between continents and ocean basins. Geol. Soc. Am. Spec. Pap. 62. Geol. Soc. Am., Boulder, CO.

Goldberg, E.D. 1976. The health of the oceans. United Nations Scientific and Cultural Organization, Paris.

Holeman, J.N. 1968. The sediment yield of major rivers of the world. Water Resour. Res. 4:737–747.

International Soil Reference and Information Center/United Nations Environment Program, 1990. World map of the status of human-induced soil degradation (Global Assessment of Soil Degradation.) Wageningen/Nairobi.

Jensen, J.H.L., and R.B. Painter. 1974. Predicting sediment yield from climate and topography. J. Hydrol. (Amsterdam) 21:371–380.

Jansson, M.B. 1988. A global survey of sediment yield. Geogr. Annale 70A:81–98.

Jing, K. 1986. The relation between soil erosion and geographical environment in the middle reach of Yellow River. Geogr. Territorial Res. 2:26–32.

Kuenen, P.H. 1950. Marine geology. John Wiley, New York.

Lal, R. 1995. Erosion crop productivity relationships for soils of Africa. Soil Sci. Soc. of Am. J. (in press).

Lal, R. 1994a. Global soil erosion by water and carbon dynamic. *In* R. Lal et al. (ed.) Soils and global change. Lewis Publishers, Chelsea, MI (in press).

Lal, R. 1994b. Sustainable land use systems and soil resilience. p. 41–67. *In* D.J. Greenland and I. Sabolcz (ed.) Soil resilience and sustainable land use. CAB Int., Wallingford, England.

L'Vovich, M.I., N.L. Bratseva, G. Ya Karasik, G.P. Medvedeva, and A.V. Meleshko. 1990. A map of contemporary erosion of the earth's surface. Mapping Sci. Remote Sensing 27:51–67.

Lopatin, G.C. 1952. Detritus in the rivers of the USSR. Zap. Vses. Geogr. Obsch. Vol. 14. Geografgiz, Moscow.

MacKenzie, F.T., and R.M. Garrels. 1966. Chemical mass balance between rivers and oceans. Am. J. Sci. 264:507–525.

Meybeck, M. 1976. Total dissolved transport by world major rivers. Hydrol. Sci. Bull. 21:265–284.

Milliman, J.D., and R.H. Meade. 1983. Worldwide delivery of river sediment to the oceans. J. Geol. 91:1–21.

Pechinov, D. 1959. Vodna eroziya i to'rd ottok. Priroda 8:49–52.

Schumm, S.A. 1963. The disparity between present rates of denudation and Orogeny. U.S. Geol. Surv. Prof. Pap. 454H:H1–H13. U.S. Geol. Surv., Washington, DC.

Schumm, S.A., and M.D. Harvey, 1982. Natural erosion in the USA. p. 15–39. *In* B.L. Schmidt (ed.) Determinants of soil loss tolerance. ASA Spec. Publ. 45. ASA and SSSA, Madison, WI.

Singh, G., R. Babu, P. Narain, L.S. Bhushan, and I.P. Abrol. 1992. Soil erosion rates in India. J. Soil Water Conserv. 47: 93–95.

United Nations Environment Program. 1990. Desertification revisited. Proc. Ad. Hoc. Consultative: Meeting on the Assessment of Desertification UNEP-Desertification Control/Project Activity, Nairobi, Kenya.

United Nations Scientific and Cultural Organization. 1978. World water balance and water resources of the earth. UNESCO studies and report in the hydrology 25. UNESCO, Paris, France.

USSR National Committee for the International Hydrological Decade. 1974. Mirovoi Vodnyi Balans i Vodnye Resursy Zemli. Gidrometeoizdat, Leningrad.

Walling, D.E. 1987. Rainfall, runoff, and erosion of the land: A global review. p. 89–117. *In* K.J. Gregory (ed.) Energetics of physical environment. John Wiley & Sons, Chichester, England.

Walling, D.E., and B.W. Webb. 1983. Patterns of sediment yield. p. 69–100. *In* K.J. Gregory (ed.) Background to palaeohydrology. John Wiley & Sons, Chichester, England.

World Resources Institute. 1992–1993. Toward sustainable development. World Resources Inst., Washington, DC.

6 Land Degradation in the World's Arid Zones

H. E. Dregne

Texas Tech University
Lubbock, Texas

Land degradation in the arid regions of the world is an enduring problem, the dimensions of which are a matter of conjecture. Lowdermilk (1953) wrote a classic article on the devastation wrought by soil erosion around the Mediterranean Sea, the Middle East, and China hundreds and thousands of years ago. He painted a dismal picture of the damage by wind and water erosion he had observed in both the humid and arid regions. Eckholm (1976) wrote an influential book on overgrazing, tree cutting, wind and water erosion, and salinization of irrigated land from a world perspective. It is an interesting, informative, and well-documented analysis of the global impact of ecological stress on land and people. Africa, three-quarters of which is dryland, has the dubious distinction of being the one continent in which food production has not kept up with population increases (World Resources Institute, 1992). Part of the reason is the overgrazing, tree cutting, and resulting wind and water erosion that plague the continent (Harrison, 1987). Expansion of cultivation into drier climatic zones and shortening or eliminating the fallow period where shifting cultivation had been practiced have contributed to the reduction in crop growth and an increase in erosion.

LAND DEGRADATION AND LAND USE

At the global scale, land degradation in the arid regions is primarily the result of four processes: vegetation degradation, water erosion, wind erosion, and salinization. Salinization largely is a problem on irrigated land, but dryland salinity (saline seepage) affects significant areas of rain-fed cropland in Australia, Canada, and the USA, and probably elsewhere.

There are other important degradation processes in the drylands, but they either are not extensive or information about their extent and severity is so scarce that meaningful estimates cannot be made. Soil compaction by machinery and livestock is an example of the latter condition. Accumulation of toxic substances such as heavy metals and persistent pesticides can have serious local effects, as can mining and tourism.

The principal land uses in the drylands are grazing, rain-fed cropping, and irrigation. Excluding hyperarid regions, rangelands cover ≈88% of the drylands,

Soil and Water Science: Key to Understanding Our Global Environment, SSSA Special Publication 41.

rain-fed cropland ≈9%, and irrigated agriculture ≈3% (Dregne & Chou, 1992). Hyperarid regions are excluded because there is no human activity of significance in that climatic zone except for irrigation in the oases. Grazing occurs in the arid, semiarid, hyperarid, and dry subhumid climatic zones, rain-fed cropping occurs dominantly in the semiarid and dry subhumid zones, and irrigation in all four dryland zones.

Rangeland degradation almost entirely is a product of the degradation of the vegetative cover due to overgrazing and woodcutting. Degradation of rain-fed croplands is largely a matter of wind and water erosion. Salinization and waterlogging are the principal degradation processes on irrigated land.

GLOBAL LAND DEGRADATION ASSESSMENT

In 1991, I made an estimate of the current extent and severity of land degradation in the arid regions (Dregne & Chou, 1992). The analysis was made for the United Nations Environment Programme as part of its preparation for the 1992 United Nations Conference on Environment and Development, which was held in Brazil. Land degradation was estimated for 100 countries that were shown to have arid lands on a map prepared by UNESCO in 1977 (UNESCO, 1979). The dimensions of the degradation problem were assessed by land use. Loss of potential productivity was the criterion by which severity of degradation was estimated.

As indicated in Table 6–1, four classes of severity, by land use, were used in the assessment: slight, moderate, severe, and very severe. Slight meant little or no degradation, very severe meant economically irreversible degradation.

For irrigated land, productivity loss was estimated from degree of soil salinization. For rain-fed cropland, the causes of potential yield loss were wind and water erosion. The criterion for rangeland productivity classes is range condition, which is commonly used by range scientists to evaluate productivity status. The class limits in Table 6–1 are those employed by the U.S. Soil Conservation Service. Placement of rangelands in the very severe degradation class is due to a poor range condition combined with the presence of mobile sand dunes, hummocks and blowouts, or gullies.

Information Sources

A variety of information sources was used to make the productivity loss estimates. Good data on the damage human-induced degradation has done to

Table 6–1. Criteria for land degradation classes.

	Percentage of loss of potential productivity, by class			
Land use	Slight	Moderate	Severe	Very severe
Irrigated land	0–10	10–25	25–50	50–100
Rainfed cropland	0–10	10–25	25–50	50–100
Rangeland	0–25	25–50	50–75	75–100

potential land productivity on a fairly extensive basis, as compared with experimental plots, are scarce. The only known well-documented instance is that of the semiarid and subhumid Palouse region in the U.S. Pacific Northwest. Yields of winter wheat (*Triticum aestivum* L.), the dominant crop, have dropped by ≈20% below the potential level as the result of water erosion (Papendick et al., 1985). In the hilly Palouse, yield potential has declined on the crests and upper slopes, but not on the lower slopes where the topsoil remained deep. Actual yields during the past 50 or more years have risen, due to improved technology, but they will never again reach the level they would have if there had not been excessive erosion.

A global survey of research on the impact of water erosion on soil productivity revealed a scarcity of experimental data on the relation (Stocking & Peake, 1985). Of the research reports identified by Stocking and Peake, nearly 60% of the 195 cited were from the USA, most published after 1980. Few developing countries had any research cited. Only one wind erosion study was listed, and it did not measure crop yield. Obviously, there are large gaps in a global picture of the erosion-soil productivity relation.

Large numbers of qualitative assessments of rangeland productivity have been made in the USA and Australia, especially. Range condition evaluations are made routinely by U.S. Soil Conservation Service and Forest Service personnel. Even then, there is considerable difference of opinion among range scientists on how much damage overgrazing has done in the western states. In the developing countries where most of the world's rangelands lie, there is little qualitative and very little quantitative information on range condition.

Most of the research on the relation of soil salinity to crop yield has been done at the U.S. Salinity Laboratory (Bernstein, 1964; Maas & Hoffman, 1976). Salinity is the major threat to irrigated agriculture, and many qualitative assessments have been made of the severity of the problem. Among the complexities of salinity damage assessment is the often-variable level of soil salinity during the year in both the horizontal and vertical distribution of soluble salt.

Numerous estimates have been made of the income lost by agriculture from the effects of land degradation. These estimates are, presumably, based upon crop yield reductions. They have been made at the local and national level for erosion (Aveyard, 1988; Bishop & Allen, 1989; Bojo, 1989; Burt, 1981; Davis & Condra, 1985; Miranowski, 1984; Science Council of Canada, 1986; Veloz et al., 1985), salinization (Aveyard, 1988; Dumsday & Oram, 1990; El-Ashry et al., 1985; Science Council of Canada, 1986), soil compaction (Aveyard, 1988; Science Council of Canada, 1986), and overgrazing (Wilcox & Thomas, 1990). The economic impact of human-induced soil compaction has received much more attention in Canada and the Australian state of New South Wales than in other countries with arid regions. It is a soil degradation problem that deserves more attention than it has received. Fortunately, its effects are reversible.

With informed opinion playing such an important role in this land degradation assessment, there obviously is much room for disagreement on the estimates. Because there are no accepted standards against which to compare this assessment, it is impossible to calculate the degree of error. My principal concern is with the accuracy of the wind erosion soil damage estimate.

STATUS OF LAND DEGRADATION

Detailed results of the 1991 assessment of land degradation in arid regions for each of 100 countries are presented in Dregne and Chou (1992). Table 6–2 is a brief summary of the global estimates of the current status of land degradation.

By continents, Asia and North America have the highest level of degraded (salinized) irrigated land, Africa, Asia, and Europe are highest in rain-fed cropland degradation (erosion), and all continents except Australia have high levels of rangeland degradation (overgrazing).

My second guessing of the degradation numbers leads me to believe that the figure for irrigated land degradation is about right, the number for rainfed cropland is twice or more what it should be, and the number for rangeland should be closer to 85%. My error in putting rainfed cropland degradation at nearly 50% is due to crediting wind erosion with doing more damage to soil productivity than it really does. And it now seems to me that there are few rangelands that have not been at least moderately degraded during the last several hundred years.

Wind Erosion

Wind erosion damage to crops is primarily a matter of sand blasting leaves and soft stems, either reducing yields if the land is not replanted or shortening the effective growing season if replanting is done. Sand blasting, however, has little or no effect on soil productivity. Burying plants or exposing the roots are other ways in which wind erosion reduces crop yields. Additionally, the wind can blow away horizons above an indurated carbonate layer, making the land useless for crop production, as has occurred in the southern High Plains. And removal of fertilizer elements when fine material is blown away is a well-known problem.

Granted that wind erosion can reduce yields and soil productivity, the question remains of how severe the economic impact is. During the disastrous Dust Bowl drought years of the 1930s, wind erosion was said to have permanently ruined 40 million acres (16 million hectares) of Great Plains soils and badly damaged 200 million acres (80 million hectares) (Hurt, 1981). When the rains returned in the late 1930s and the 1940s, yields also rebounded, producing bumper crops (Hurt, 1981). Whatever permanent damage was done to soils—and there certainly was some—it had little effect on crop yields. The possibility always exists that improved technology has offset the long-term adverse effects of wind erosion; however, there is no known evidence of that. Yields still depend heavily on how much it rains and on the seasonal distribution of that rain.

Table 6–2. Global land degradation in drylands, by land use.†

Land use	Total area 1000 ha	Degraded‡ %
Irrigated land	145000	30
Rainfed cropland	457000	47
Rangeland	4556000	73
World	5160000	70

†Source: Dregne and Chou, 1992.
‡Includes moderate, severe, and very severe degradation (all except slight degradation).

The only known study on wind erosion costs was carried out in New Mexico for the Soil Conservation Service (Davis & Condra, 1985; Huszar, 1985). That study showed that off-site effects due to blowing sand and dust were ≈45 times greater than on-site effects due to reduced production and increased operating costs. Davis and Condra (1985) concluded that the cost of productivity loss due to soil loss from wind erosion was generally considered small. New Mexico has the second highest wind erosion rate, behind Texas, of the 10 Great Plains states (U.S. Department of Agriculture, 1981).

REVERSIBILITY

Most rangeland vegetation degradation is reversible (Stoddart et al., 1975). The exception is when vegetation degradation is accompanied by permanent changes in soil properties caused by erosion and soil trampling by livestock. For example, very severely degraded rangeland is considered to be economically irreversibly degraded. If expense is no concern, any land can be restored to its former productivity. The principal difficulty in improving rangelands is the climate hazard. In the drier climatic zones, restoration of the climax vegetation or a reasonable facsimile of it may take decades, whereas in the subhumid zones significant change may occur within a few years.

In the case of wind and water erosion, there is no restoration of the original soil horizons over human lifetimes when erosion has caused moderate or worse changes in horizon thickness or soil physical properties. Stopping further degradation and improving physical conditions and fertility is all that can be expected. Gullies are an obvious example of an irreversible loss.

Salinization is reversible in all but a few cases. Reversing salinization can, however, be costly. Fortunately, the cost/benefit ratio usually is favorable because of the generally high productivity of irrigated land. Reversing dryland salinity is more problematic, economically.

CONCLUSIONS

Land degradation in the arid regions is a global problem, with virtually all 100 countries having, within their borders, significant drylands containing large areas of degraded land. Rangelands occupy nearly 90% of the drylands, excluding hyperarid regions, and most of it is degraded to varying degrees. Between one-quarter and one-third of the irrigated lands are believed to be salinized sufficiently to reduce crop yields. There is considerable uncertainty about the degradation status of rainfed croplands. At least one-quarter probably have experienced moderate or worse water erosion.

A major question is the severity of the impact of wind erosion on long-term soil productivity. Most people probably believe that wind erosion has damaged extensive areas in the drylands. While off-site effects undoubtedly are high, on-site damage is not so easy to evaluate.

Vegetation degradation of rangelands and salinization of irrigated land are largely reversible. It may not be economic to do so on very severely degraded rangelands and irrigated land, however. Restoring eroded soils to their original condition is impossible in human lifetimes, but most degraded soils probably can be improved greatly under good management.

REFERENCES

Aveyard, J.M. 1988. Land degradation: Changing attitudes—why? J. Soil Conserv. New South Wales 44:46–51.

Bernstein, L. 1964. Salt tolerance of plants. USDA Inform. Bull. 283. USDA, Washington, DC.

Bishop, J., and J. Allen. 1989. The on-site costs of soil erosion in Mali. Environment Working Pap. 21. Environment Department, World Bank, Washington, DC.

Bojo, J. 1989. Case study: Benefit-cost analysis of soil conservation in Maphutseng, Lesotho. p. 251–285. *In* J.A. Dixon et al. (ed.) The economics of dryland management. Earthscan Publ. Ltd., London.

Burt, O.R. 1981. Farm level economics of soil conservation in the Palouse area of the Northwest. Am. J. Agric. Econ. 63:83–92.

Davis, B., and G.D. Condra. 1985. The on-site cost of wind erosion in New Mexico. Dep. of Agric. Econ., Texas Tech Univ., Lubbock.

Dregne, H.E., and N.-T. Chou. 1992. Global desertification dimensions and costs. p. 249–282. *In* H.E. Dregne (ed.) Degradation and restoration of arid lands. Texas Tech Univ. Press, Lubbock.

Dumsday, R.G., and D.A. Oram. 1990. Economics of dryland salinity control in the Murray River Basin, northern Victoria (Australia). p. 215–240. *In* J.A. Dixon et al. (ed.) Dryland management: Economic case studies. Earthscan Publ., Ltd., London.

Eckholm, E.P. 1976. Losing ground. W.W. Norton & Company, New York.

El-Ashry, M.T., J. van Schilfgaarde, and S. Schiffman. 1985. Salinity pollution from irrigated agriculture. J. Soil Water Conserv. 40:48–52.

Harrison, P. 1987. The greening of Africa. Int. Inst. for Environ. and Develop.–Earthscan, Washington, DC.

Hurt, R.D. 1981. The Dust Bowl. Nelson-Hall, Chicago.

Huszar, P.C. 1985. Off-site economic costs of wind erosion in New Mexico. Dep. of Agric. and Resour. Econ., Colorado State Univ. ANRE Res. Rep. AR:85-3. Colorado State Univ., Fort Collins.

Lowdermilk, W.C. 1953. Conquest of the land through seven thousand years. USDA Agric. Inform. Bull. 99. USDA-SCS, Washington, DC.

Maas, E.V., and G.J. Hoffman. 1976. Crop salt tolerance: Evaluation of existing data. p. 187–198. *In* H.E. Dregne (ed.) Managing saline water for irrigation. Texas Tech Univ., Lubbock.

Miranowski, J.A. 1984. Impacts of productivity loss on crop production and management in a dynamic economic model. Am. J. Agric. Econ. 66:61–71.

Papendick, R.I., D.L. Young, D.K. McCool, and H.A. Krauss. 1985. Regional effects of soil erosion on crop productivity—The Palouse area of the Pacific Northwest. p. 305–320. *In* R.F. Follett and B.A. Stewart (ed.) Soil erosion and crop productivity. ASA, CSSA, and SSSA, Madison, WI.

Science Council of Canada. 1986. A growing concern: Soil degradation in Canada. Ottawa, Canada.

Stocking, M., and L. Peake. 1985. Erosion-induced loss in soil productivity: Trends in research and international cooperation. Univ. of East Anglia Overseas Development Group, Norwich, England.

Stoddart, L.A., A.D. Smith, and T.W. Box. 1975. Range management. 3rd ed. McGraw-Hill Book Co., New York.

United Nations Educational, Scientific, and Cultural Organization. 1979. Map of the world distribution of arid regions. Explanatory Note. UNESCO, Paris.

U.S. Department of Agriculture. 1981. 1980 Appraisal. Part 1. Soil, water, and related resources in the United States: Status, conditions, and trends. USDA, Washington, DC.

Veloz, A., D. Southgate, F. Hitzhusen, and R. MacGregor. 1985. The economics of erosion control in a subtropical watershed: A Dominican case. Land Econ. 61(2):145–155.

Wilcox, D.G., and J.F. Thomas. 1990. The Fitzroy Valley Regeneration Project in Western Australia. p. 116–137. *In* J.A. Dixon et al. (ed.) Dryland management: Economic case studies. Earthscan Publ., Ltd., London.

World Resources Institute. 1992. World Resources 1992–93. World Resour. Inst., Washington, DC.

7 Agriculture in a Changing Global Environment

Cynthia Rosenzweig
Columbia University and
Goddard Institute for Space Studies
New York, New York

"A glimpse of earth from space should be sufficient to restore the true perspective. It shows the planet whole, without political or tribal boundaries. How beautiful, how colorful, how delicate is this ball of lapping waters, floating continents, and swirling clouds gliding in a thin veil of air. And how small, unique, and solitary is this one and only home of ours. We must listen to its signals of distress, for it is our parent and we are all its dependent children." These sentences, taken from Daniel Hillel's book, *Out of the Earth* (Hillel, 1992), describe the planet as seen from the Apollo mission (Fig. 7–1), a perspective that is enlarging the scope of environmental research.

Environmental scientists are now engaged in a quest to understand the processes that shape the global environment. The quest is primarily driven by the realization that modern civilization has been modifying the composition of the atmosphere that envelops our earth, the very medium within which we live and breathe. The consequences of this change now threaten the stability of the earth's climate.

GLOBAL CLIMATE CHANGE

Burning of fossil fuels and eradication of forests have already raised the atmospheric concentration of CO_2 by some 25% since the industrial revolution, and that rise continues at a rate of $\approx$0.5% yr^{-1} (Fig. 7–2) (Houghton et al., 1990, 1992). Despite its seemingly minute concentration (only 0.035%), CO_2 inhibits the escape of longwave (thermal) radiation emitted by the earth throughout the entire atmosphere, a process that leads to eventual surface warming and associated feedback effects.

Other gases that are present in even smaller concentrations, but that similarly tend to trap heat include methane, nitrous oxide, and a family of particularly insidious synthetic gases, the chlorofluorocarbons, insinuated into the atmosphere only in the last 50 yr. These radiatively active gases, which together with CO_2 are called the greenhouse gases, are also building up in the atmosphere, mostly as a result of human activity (Fig. 7–2) (Houghton et al.,

Soil and Water Science: Key to Understanding Our Global Environment, SSSA Special Publication 41.

Fig. 7–1. Planet Earth viewed from the Apollo Mission (Source: NASA photo, NASA, Washington, DC).

1990, 1992). Other changing atmospheric factors that affect climate include ozone, solar irradiance, tropospheric aerosols, and stratospheric aerosols (Hansen et al., 1993).

So this much is clear: if the accumulation of greenhouse gases continues unabated, it is bound sooner or later to have a warming effect on the earth's climate. Such a warming trend is likely to affect the regional patterns of precipitation and evaporation, indeed the entire array of meteorological, hydrological, ecological, and agricultural relationships.

Beyond this truism, however, lie great uncertainties: how much warming will occur, at what rate, and according to what geographical and seasonal pattern? And just what will be the consequences to the agricultural productivity of different countries and regions? Will some nations benefit, while others suffer, and who might the winners and losers be? Finally, there are the practical questions: What can and should be done in timely fashion by individual countries and by the international community as a whole to avert potential damages to

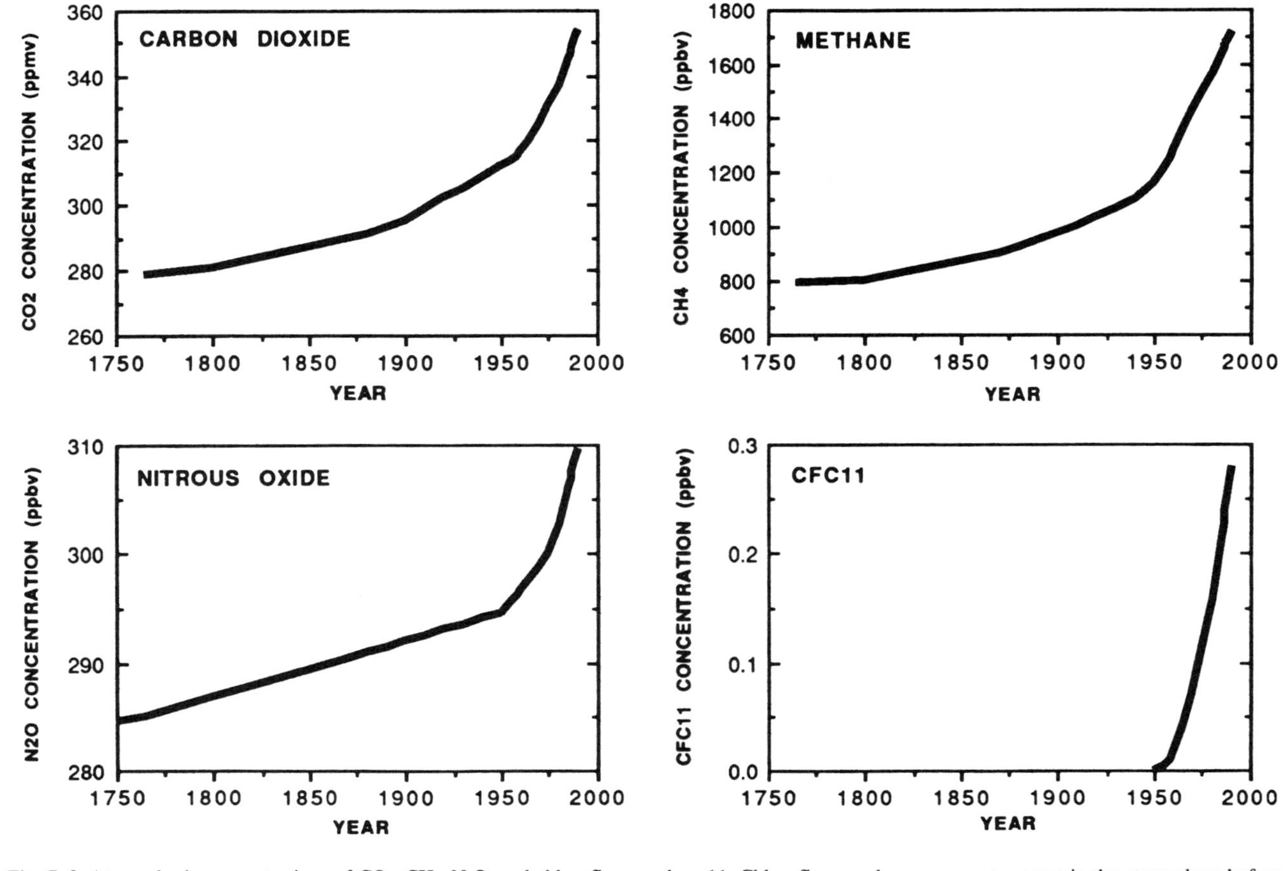

Fig. 7–2. Atmospheric concentrations of CO_2, CH_4, N_2O, and chlorofluorocarbon-11. Chlorofluorocarbons were not present in the atmosphere before the 1930s (Source: Houghton et al., 1990).

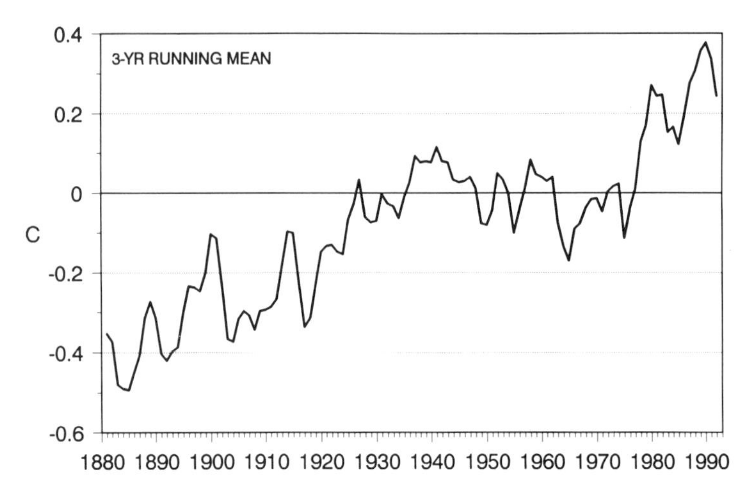

Fig. 7–3. Observed global surface air temperature, relative to 1951–1980 (Source: Wilson and Hansen, personal communication, updated version of Hansen & Lebedeff, 1987).

life-support systems? And, to the extent that such damages are not completely avoidable, what can be done to minimize or overcome them? Upon our ability to answer such questions may rest the fate of natural and human-controlled ecosystems.

As yet, the evidence regarding the link between observed temperature trends and the enhanced greenhouse effect is ambiguous and nothing is absolutely proven (Houghton et al., 1990, 1992). Global mean surface air temperature has risen by 0.3 to 0.6°C during the past century (Fig. 7–3) (e.g., Hansen & Lebedeff, 1988), and some observers have suggested that the predicted enhanced greenhouse warming may have already begun (Hansen, 1988). While the amount of warming over the long-term record is consistent with the effect expected from the trace gases already emitted into the atmosphere (Houghton et al., 1990, 1992), the signal is still not clearly discernible from the noise.

Recent observations suggest that diurnal minimum (nighttime) temperatures are rising more than maximum (daytime) temperatures (Karl et al., 1991), possibly related to increased cloudiness, polluting S aerosols, or enhanced greenhouse effect at night (Houghton et al., 1992). These differential temperature effects can be important for crop responses, such as nighttime respiration rates, winter kill, and vernalization. Recent short-term cooling of 0.3°C in 1992 has been linked to stratospheric volcanic aerosols from the 1991 eruption of Mt. Pinatubo (Hansen et al., 1993).

In the absence of actual proof, scientists have been employing mathematical models of the climate system (involving the atmosphere, the oceans, and land masses) to assess the processes known to occur and their possible interactions, and to project their potential effects into the future. The strength of these models, known as global climate models (GCMs), is that they integrate our best knowledge, allowing calculation of complex feedback mechanisms; their weakness is that they also embody our ignorance. The GCM deficiencies result, in part, from incompletely understood ocean circulation patterns, lack of knowledge concerning the formation and feedback effects of clouds (whether positive or negative), simplistically formulated hydrological processes, and coarse spatial resolution. So GCM results are still tentative and should not be accepted uncritically.

Nevertheless, it is important to consider their predictions and limitations, while continuing to look for the emerging empirical evidence of changing climate. Since such changes are likely to impact agriculture, energy use, and other aspects of the economies of both developed and developing countries, it is important that we make the effort to improve our understanding of global climate change, study its potential impacts, and develop policies for dealing with it.

AGRICULTURAL EMISSIONS OF GREENHOUSE GASES

The role of climate as a primary determinant of agriculture has long been recognized. It is only in the last decade, however, that agriculture's reciprocal effect on climate change has come to light. Clearing forests for fields, burning crop residues, submerging land in rice (*Oryza sativa* L.) paddies, raising large herds of ruminants, and fertilizing with N, all release greenhouse gases to the atmosphere. The main gases emitted are CO_2, CH_4, and N_20. Emissions from agricultural sources account for ≈15% of total anthropogenic greenhouse gas emissions and land-use change (often for agricultural purposes) contributes another 8% (Fig. 7–4) (Lashof & Tirpak, 1989). Agriculture ranks third after energy

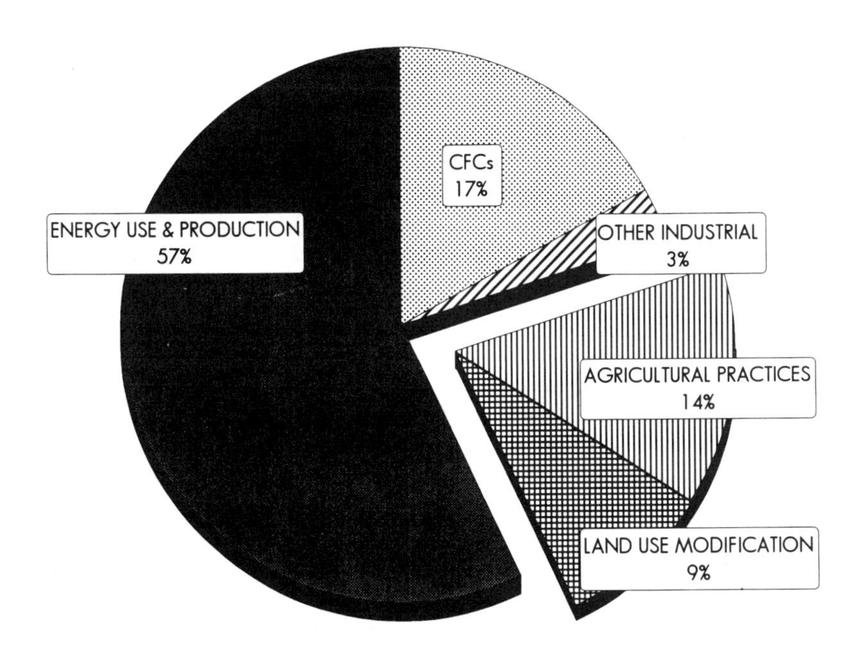

Fig. 7–4. Present contributions of anthropogenic activities to global warming (Source: Lashof and Tirpak, 1989).

consumption and chlorofluorocarbon production as a contributor to the enhanced greenhouse effect.

Carbon Dioxide

Carbon in various forms (e.g., CO_2, carbonates, or organic compounds) is cycled between the atmosphere, oceans, land biota, and marine biota on short time scales and into sediment and rocks on geological time scales. Through the process of photosynthesis, vegetation removes about 100 Gt C from the atmosphere annually. About one-half the C thus synthesized is later respired back to the atmosphere, while the other one-half is incorporated into plant tissue and (from plant residues) into soil organic matter. On a global basis, soil is second only to the ocean in the amount of C it contains, although there are large uncertainties in the range of total soil C (1000–2500 Pg C) (Bouwman, 1990).

Agricultural practices manipulate the vegetation and soil carbon reservoirs (Jackson, 1993). When land supporting a natural ecosystem is converted to agricultural use, C stored in plant biomass is burned or decomposed and the organic matter in the soil is oxidized. The net end product is CO_2, because forests store 20 to 100 times more C per unit area than does cropland (Houghton et al., 1983). Biomass burning also results in emissions of CH_4, N_2O, NO_x, and CO; N_2O is also released from exposed soil.

Over time, clearing land to facilitate food production has probably made the largest contribution to the greenhouse gas buildup of any agricultural activity (Burke & Lashof, 1990). Agricultural expansion of cropland and pasture may well have been responsible for the initial rise in CO_2 emissions noticed since the beginning of the 19th century; fossil fuel burning probably did not begin to dominate emissions until the 1950s. Since 1850, conversion to cropland has replaced ≈15% of the world forest area (Houghton et al., 1983) and approximately one-third of the earth's land surface is now devoted to agricultural cropland and pasture (Food and Agriculture Organization, 1986). At present, ≈11 million hectares of tropical forest are converted each year, while only ≈1 million hectares are reforested (Food and Agriculture Organization, 1986). The annual net flux of C to the atmosphere from deforestation is estimated to be 2 Gt C (Houghton et al., 1990).

Methane

Methane is the second most important greenhouse gas after CO_2 in terms of radiative forcing. Its current atmospheric concentration is 1.72 ppmv and it is increasing at a rate of ≈0.75% per year (Houghton et al., 1992). On a molecular basis, CH_4 is ≈20 times more effective than CO_2 in affecting climate, given current atmospheric concentrations (Houghton et al., 1990). Agriculture is the dominant anthropogenic source of methane (Houghton et al., 1990, 1992). The major agricultural sources that release CH_4 are wetland rice, ruminant animals, and biomass burning.

Anaerobic decomposition of biomass in flooded rice fields produces CH_4, which is transported to the atmosphere through the rice plants and standing water (Bouwman, 1991). Methane emissions from rice paddies vary with soil type, nutrients, redox potential, pH, and temperature, and with agronomic practices such as irrigation and water management practices, cropping systems, fertilization, and additions of manure or rice straw (Yagi & Minami, 1991; Bouwman, 1991; Lindau et al., 1993).

Ruminant animals, cattle (*Bos* sp.), sheep (*Ovis aries*), goats (*Capra hircus*), camels (*Camelus* sp.), and buffalo (*Babalus babalus*), are primarily responsible for CH_4 production via the process of enteric fermentation in the breakdown of cellulose. Horses (*Equus caballus*), pigs (*Sus scrofa*), and humans (!), nonruminants, also emit methane. Of the total emissions from enteric fermentation, cattle account for ≈57%, dairy cows 19%, and sheep 10% (Burke & Lashof, 1990). Dairy and beef cattle are most important both because of high emission rate per animal (≈55 kg animal^{-1} yr^{-1}) and high population (>1.25 billion in the late 1980s). Methane production varies with feed quality and amount, animal age, weight, genetics, activity, and enteric ecology. Agricultural methane releases are also associated with tropical forest clearing for crop and livestock production and with burning of crop residues.

Nitrous Oxide

Nitrous oxide is present in the atmosphere in even smaller quantities than CO_2 and CH_4. The current concentration is 310 ppbv compared with 355 ppmv for CO_2 and 1.72 ppmv for CH_4 (Houghton et al., 1992). The annual growth rate of N_2O is 0.2 to 0.3% per year (Houghton et al., 1992). The radiative forcing of the N_2O molecule is ≈230 times greater than that of CO_2. Besides being a greenhouse gas, N_2O also plays an important role in the atmosphere because it produces NO_x as it breaks up in the stratosphere, leading to depletion of stratospheric ozone.

There are large uncertainties in the estimates of N_2O emissions both from natural sources and human activities, due to lack of measurements of N_2O fluxes, the complexity of the biogeochemical interactions, and the heterogeneity of the land surface. Emissions from soils appear to dominate the N_2O budget (Houghton et al., 1992). Nitrous oxide is released naturally from soils via the microbial metabolic processes of nitrification-denitrification and NO_3 assimilation-dissimilation. Human activities connected to N_2O emissions are the use of nitrogenous fertilizers, biomass burning, land-use changes, and fossil-fuel combustion. Of these, agriculture plays a dominant role in fertilizer use, biomass burning and land-use change. Significant amounts of N_2O are emitted from agricultural soils, particularly those fertilized with organic manures or those having a high organic matter content (Sahrawat & Keeney, 1986). Emissions of N_2O from fertilizer applications are estimated to be ≈10% of total emissions (Burke & Lashof, 1990). Biomass burning releases a relatively small amount of N_2O to the atmosphere; this flux is now estimated at only 0.02 to 0.2 Tg (1 Tg = 10^{12} g) of N per year (Houghton et al., 1992). To the extent that agriculture uses fossil fuels for energy, it also contributes to the fossil-fuel N_2O source.

Emissions of greenhouse gases from agricultural sources are likely to increase in the future, given the need for expansion of production to feed our growing population. This poses a challenge to agricultural researchers to continue to improve yields, while holding down emissions. Some directions are indicated: reductions of rate of land-clearing and biomass burning in the tropics; management of rice paddies and livestock to reduce methane emissions; and improved fertilizer-use efficiency. Much research, however, is still needed to understand the processes by which greenhouse gases are emitted from different agricultural practices and systems and how to reduce the emissions. Reductions require that knowledge of effective techniques be disseminated and put into practice. It is likely that reductions in some gases will prove to be more easily achievable, so that efforts may be concentrated on greater reductions of some gases rather than others; however, such strategies will vary by region. Finally, if climate is changing at the same time, complex interactions are likely to ensue.

POTENTIAL IMPACTS OF CLIMATE CHANGE ON AGRICULTURE

In the coming decades, global agriculture faces the prospect of a changing climate, as well as the known challenge of feeding the world's population as it grows from 5 to 10 billion people (International Bank for Reconstruction and Development, World Bank, 1990). Global warming is predicted to bring changes in the thermal and hydrologic regimes of entire regions. The nature of such changes and their biophysical and socioeconomic consequences is the subject of the research field known as climate change impacts (Tegart et al., 1990). Agricultural impacts research aims to analyze and elucidate the predictable changes and to evaluate what practical adaptations might be undertaken to prevent or overcome any possible adverse impacts on our ability to feed the world. Several complementary disciplines are needed to implement the complex task of assessing potential climate change impacts on global agriculture. These include the relevant aspects of atmospheric science, hydrology, soil science, crop physiology, and resource economics (Fig. 7–5).

Climate variables can have significant impacts on the quantity, quality, and regional pattern of agricultural production, because heat, light, and water are major biophysical drivers of crop growth. Suboptimal levels of these factors can bring severe consequences. The Dust Bowl drought of the 1930s brought nearly 200 000 farm bankruptcies to the southern Great Plains, and the recent drought of 1988 in the Midwest led to a 30% reduction in U.S. corn production (Rosenzweig & Hillel, 1993). Above normal temperatures accompanied both of these droughts. Higher temperatures in general accelerate the phenology of annual cereal crops, resulting in hastened maturation and reduced yield potential (e.g., Butterfield & Morison, 1992). On the other hand, warmer and longer growing seasons in regions where crops are currently limited by cold, but not by paucity of moisture, may enjoy increased productivity in a warmer world.

If atmospheric CO_2 accumulations were occurring without concomitant changes in temperature and water regimes, it might indeed be a blessing for crop production. Atmospheric CO_2 is an essential ingredient of the basic process of photosynthesis by which that gas combines with soil-derived water with the infusion of sunlight to create carbohydrates and, ultimately, food for humans and other animals. Plants growing in air with higher CO_2 exhibit increased rates of photosynthesis because the absorption of CO_2 is facilitated by the increased con-

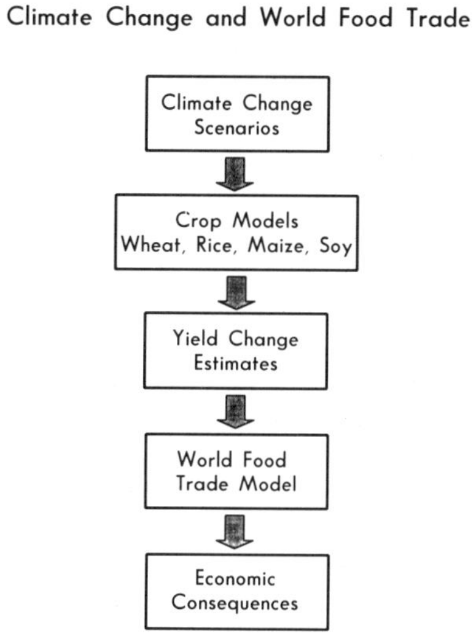

Climate Change and World Food Trade

Climate Change Scenarios

Crop Models Wheat, Rice, Maize, Soy

Yield Change Estimates

World Food Trade Model

Economic Consequences

Fig. 7–5. Design of a climate change impact study on crop yields and world food supply (Source: Rosenzweig et al., 1993).

centration gradient between the external atmosphere and the air spaces inside the leaves (Acock & Allen, 1985).

Plants also exhibit partial closure of stomates in the presence of elevated CO_2 levels, thereby reducing transpiration on a per-unit leaf area basis (Morison & Gifford, 1984). Total crop evapotranspiration, however, may not change significantly because the greater foliar growth due to the higher rates of photosynthesis partially compensates for the reduced transpiration per unit of leaf area (Allen et al., 1985). A net improvement may nevertheless result in crop water-use efficiency, defined as the ratio between crop biomass accumulation and the amount of water used in evapotranspiration. Considering these interactive processes in combination is a complex task, requiring a quantitative dynamic assessment of the relative magnitudes of both physiological and climatic changes.

How climate change affects agriculture also depends on how the biophysical field-level effects (including those on soil fertility and pests) lead to changes in the socio-economic sphere within which farmers operate (Adams et al., 1990). Decisions made by individual farmers in the face of changing climatic regimes will underlie society's collective response to climate change. These decisions involve cropping and irrigation systems, risk management, and ultimately the successes or failures of individual farms and, by extension, of entire regions and even of national economies highly dependent on agricultural products. Their effects may cascade into such areas as energy and machinery requirements, irrigation water use, storage and transportation facilities, and regional markets.

National farm policy is a critical determinant of the future of U.S. agriculture, especially in light of potential climate change. U.S. farm subsidies (currently costing ≈$10 billion per year) must be examined to see if they serve to help or hinder necessary adaptation to the eventuality of a changing climate. An important policy consideration is the assessment of risk due to weather anomalies. The drought of 1988, for example, cost $3 billion in direct payments to farmers. If drought frequency increases, the need for such emergency subsidies will also increase.

Beyond national boundaries, global patterns of supply and demand may change in far-reaching ways (Rosenzweig & Parry, 1994). Because of the interdependence of the world food system, the consequences of climate change for agriculture in the USA depends, to a large extent, on what happens elsewhere, and vice versa. Vulnerability of marginal or food-deficient regions to climate change may create markets for U.S. grain, but improved comparative advantage in more productive areas may limit those markets.

Although some countries in the temperate zones may reap some benefit, many countries in the tropical and subtropical zones appear to be more vulnerable (Rosenzweig & Parry, 1994). Particular hazards are increased flooding of low-lying areas, the increased frequency and severity of drought in semiarid areas, and potential decreases in crop production in developing countries (Fig. 7–6). International trade policy issues, such as the movement to lower agricultural trade barriers, and food security planning will be crucial in climate change response strategies.

Agriculture is no longer perceived as a benign or even neutral influence on the environment. Not only does agricultural development replace natural ecosystems with artificially managed ones wherever it takes place, but it also affects its

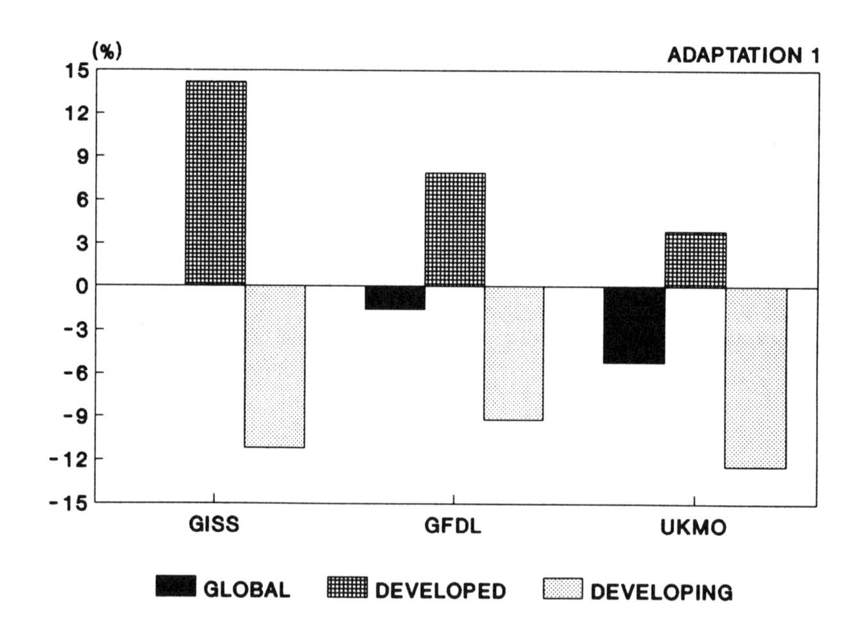

Fig. 7–6. Change in cereal production (with physiological CO_2 effects on crop growth and water use) in 2060 for three GCM climate change scenarios (GISS, Goddard Inst. for Space Studies; GFDL, Geophysical Fluid Dynamics Lab.; UKMO, United Kingdom Met. Office). Adaptation level 1 implies minor changes to existing agricultural systems. (Source: Rosenzweig and Parry, 1994).

surroundings by its increased rates of erosion and runoff, and by its release of fertilizer and pesticide residues into surface waters and groundwaters (Hillel, 1992). Moreover, as a result of potentially large climate changes, crop and livestock patterns may shift, invading regions now primarily covered by forest and other types of less intensively managed ecosystems. Such interactions among the changing climate, agriculture, and the environment will reverberate throughout the world food system, affecting rates of soil erosion, fates of agricultural chemicals, and wildlife habitats. Of equal significance, they may also intensify competition for water resources.

CONCLUSION

A host of complex interactions between and among biophysical and socio-economic factors of climate change will impact agriculture. And agricultural practices will impact climate through the emission of greenhouse gases. Methods of climate change prediction, estimation of physiological and climatic crop responses, and analysis of socio-economic consequences are improving. The goals of climate change impact research are to define more clearly the ranges of possible consequences, to identify the critical thresholds of our agricultural system, and to engender flexibility in society's preparatory and responsive actions. These tasks are complicated by the ongoing degradation of land and water resources through prolonged abuse and the potential for multiple environmental stresses on crop growth and production, among them increases in tropospheric ozone and increased ultra violet-B radiation.

There is a growing emphasis on the need for integrated assessment that considers both biophysical and socioeconomic consequences of climate change over time. Interactions of the agricultural system with other sectors: energy, water resources, and natural ecosystems, and their potential feedbacks on trace gas emissions are also beginning to be studied. Since future agricultural practices are likely to be different from today's practices, estimates of these improvements should be factored into projections. For example, conventional plant breeding and biotechnology are each expected to improve crop yields in the next century. Other probable future changes in agriculture include higher fossil fuel prices and greater use of crops for fuel, fiber, and other nonfood products.

Both national and international research organizations should encourage the development of new approaches likely to be effective in preparing for climate change. Agricultural research would benefit from increased attention to both macroclimate and microclimate in all experiments and variety trials. Considering the vulnerability of agricultural production to the occurrence of climate extremes, research should be directed to determine what are the heat-tolerance limits of currently grown and of alternative crops and varieties. Adjustments may also include modification of agronomic practices (e.g., the timing and modes of tillage and planting), adoption of crops known to be heat- and drought-resistant, increased efficiency of irrigation and water conservation, and improved pest management. Even aside from global climate change considerations, such adjustments are worthy of being implemented.

REFERENCES

Acock, B., and L.H. Allen, Jr. 1985. Crop responses to elevated CO_2 concentrations. p. 53–97. *In* B.R. Strain and J.D. Cure (ed.) Direct effects of increasing CO_2 on vegetation. DOE/ER-0238. U.S. Department of Energy, Washington, DC.

Adams, R.M., C. Rosenzweig, R.M. Peart, J.T. Ritchie, B.A. McCarl, J.D. Glyer, R.B. Curry, J.W. Jones, K.J. Boote, and L.H. Allen, Jr. 1990. Global climate change and U.S. agriculture. Nature (London) 345:219–222.

Allen, L.H., Jr., P. Jones, and J.W. Jones. 1985. Rising atmospheric CO_2 and evapotranspiration. p. 13–27. *In* Advances in evapotranspiration. Proc. of the Natl. Conf. on Advances in Evapotranspiration, St. Joseph, MI, 16–17 Dec. 1985. ASAE, St. Joseph, MI.

Bouwman, A.F. 1990. Soils and the greenhouse effect. John Wiley & Sons, New York.

Bouwman, A.F. 1991. Agronomic aspects of wetland rice cultivation and associated methane emissions. Biogeochemistry 15:65–88.

Burke, L.M., and D.A. Lashof. 1990. Greenhouse gas emissions related to agriculture and land-use practices. p 27–43. *In* B.A. Kimball et al. (ed.) Impact of carbon dioxide, trace gases, and climate change on global agriculture. ASA Spec. Publ. 53. ASA, CSSA, and SSSA, Madison, WI.

Butterfield, R.E., and J.I.L. Morison. 1992. Modeling the impact of climatic warming on winter cereal development. Agric. For. Meteorol. 62:241–261.

Food and Agriculture Organization. 1986. FAO production yearbook. Vol. 39. FAO, Rome.

Hansen, J.E. 1988. The greenhouse effect: Impacts on current global temperature and regional heat waves. Testimony to U.S. Senate Committee on Energy and Natural Resources. 23 June 1988. U.S. Gov. Print. Office, Washington, DC.

Hansen, J., A. Lacis, R. Ruedy, M. Sato, and H. Wilson. 1993. How sensitive is the world's climate? Res. Exploration 9(2):142–158.

Hansen, J., and S. Lebedeff. 1987. Global trends of measured surface air temperature. J. Geophys. Res. 92:13345–13372.

Hansen, J., and S. Lebedeff. 1988. Global surface air temperatures: Update through 1987. Geophys. Res. Lett. 15:323–326.

Hillel, D. 1992. Out of the earth: Civilization and the life of the soil. Univ. of California Press, Berkeley.

Houghton, J.T., B.A. Callander, and S.K. Varney (ed.). 1992. Climate change 1992: The supplementary report to the IPCC scientific assessment. World Meteorological Organization and United Nations Environment Programme. Cambridge Univ. Press, Cambridge.

Houghton, R., J. Hobbie, and J. Melillo. 1983. Changes in the carbon content of terrestrial biota and soils between 1860 and 1980: A net release of CO_2 to the atmosphere. Ecol. Monogr. 53:235–262.

Houghton, J.T., G.J. Jenkins, and J.J. Ephraums (ed.). 1990. Climate change: The IPCC scientific assessment. World Meteorological Organization and United Nations Environment Programme. Cambridge Univ. Press, Cambridge.

International Bank for Reconstruction and Development–World Bank. 1990. World Population Projections. Johns Hopkins Univ. Press, Baltimore.

Jackson, R.B., IV. 1993. Greenhouse gases and agriculture. p. 417–444. *In* R.A. Geyer (ed). A global warming forum: Scientific, economic, and legal overview. CRC Press, Boca Raton.

Karl, T.R., G. Kukla, V.N. Razuvayev, M.J. Changery, R.G. Quayle, R.R. Heim, Jr., D.R. Easterling, and C.B. Fu. 1991. Global warming: Evidence for asymmetric diurnal temperature change. Geophys. Res. Lett. 18:2253–2256.

Lashof, D.A., and D.A. Tirpak (ed.). 1989. Policy options for stabilizing global climate. Rep. to Congress. 21P-2003.1. U.S. Environmental Protection Agency Office of Policy, Planning and Evaluation, Washington, DC.

Lindau, C.W., W.H. Patrick, Jr., and R.D. DeLaune. 1993. Factors affecting methane production in flooded rice soils. p. 157–165. *In* L.A. Harper et al. (ed.) Agricultural ecosystem effects on trace gases and global climate change. ASA Spec. Publ. 55. ASA, CSSA, and SSSA, Madison, WI.

Morison, J.I.L., and R.M. Gifford. 1984. Plant growth and water use with limited water supply in high CO_2 concentrations: I. Leaf area, water use and transpiration. Aust. J. Plant Physiol. 11:361–374.

Rosenzweig, C., and D. Hillel. 1993. The Dust Bowl of the 1930s: Analog of greenhouse effect in the Great Plains? J. Environ. Qual. 22:9–22.

Rosenzweig, C., M.L. Parry, G. Fischer, and K. Frohberg. 1993. Climate change and world food supply. Res. Rep. 3. Environmental Change Unit, Univ. of Oxford, Oxford.

Rosenzweig, C., and M.L. Parry. 1994. Potential impact of climate change on world food supply. Nature (London) 367:133–138.

Sahrawat, K.L., and D.R. Keeney. 1986. Nitrous oxide emission from soils. Adv. Soil Sci. 4:103–148.

Tegart, W.J. McG., G.W. Sheldon, and D.C. Griffiths (ed.). 1990. Climate change: The IPCC impacts assessment. World Meteorological Organization and United Nations Environment Programme. Australian Gov. Publ. Serv., Canberra.

Yagi, K., and K. Minami. 1991. Emission and production of methane in the paddy fields of Japan. JARQ 25:165–171.

8 Role of Geopurification in Future Water Management

Herman Bouwer

U.S. Water Conservation Laboratory
Phoenix, Arizona

There is no end in sight to the increases in world population, which is predicted to double from the present 5.6 billion to ≈10 billion by the middle of the next century (State of World Population Report released in 1993 by the United Nations Population Fund). Ninety-five percent of the population increase is expected to occur in the mostly tropical and subtropical countries of the Third World, which may have ≈87% of the world's population by the year 2050. Such population growth will present tremendous pressures and challenges on the earth to provide adequate water, food, shelter, and livelihood for this mass of humanity.

There are still vast underpopulated areas in the world where soil resources outnumber water resources. On the other hand, there are also densely populated areas with plentiful water supplies but limited availability of land and soil. The problem thus primarily is distribution of people and soil and water resources. Massive migration can be expected from dry to wet, poor to rich, and rural to urban areas. Presently, already ≈100 million people live outside the countries where they were born. Cities will continue to grow and by the year 2000 there will be ≈22 megacities of >10 million people with 18 of them in the Third World. Population growth in the First World has almost stopped, except perhaps in the USA where the population may double in the next century, depending on immigration policies.

The most common limiting factor in population expansion is the availability of water resources. As a rough estimate, a renewable water supply of at least 2000 m^3 $person^{-1}$ yr^{-1} is required to provide an adequate standard of living (Postel, 1992). When the water supply drops to between 1000 and 2000 m^3 $person^{-1}$ yr^{-1}, the area is considered water stressed, and below 1000 m^3 $person^{-1}$ yr^{-1}, it is considered water scarce. Water resources planning in China is based on ≈500 m^3 $person^{-1}$ yr^{-1}. All of us are familiar with the starvation, diseases, lack of economic development, and just plain misery caused by chronic water shortages, especially in the Third World.

WATER RESOURCES MANAGEMENT

Water resources can be locally or regionally increased by cloud seeding. The significance of this practice is still debated and studied. Desalination will enable

Soil and Water Science: Key to Understanding Our Global Environment, SSSA Special Publication 41.

use of seawater in coastal areas or salty groundwater in inland areas. Desalination is expensive; i.e., about $1 per 1000 m^3 for every 10 mg L^{-1} salt removed with reverse osmosis, or about $2000 to $3000 per 1000 m^3 for seawater. For inland areas, large-scale desalination presents problems of disposal of the reject water (brine), which may have to be flash-evaporated with disposal of the salts in landfills or other designated depositories. Above all, increasing demands for water will require more intensive management of water resources and water conservation. There must be more storage and regulation of stream flow to store water during times of water surplus for use in times of water shortage. The traditional approach of building more dams may not be the best solution because the world is running out of good dam sites and dams have a finite useful life, structurally as well as due to sediment accumulation. Also, environmental and social opposition to dams is growing. Thus, there must be more underground storage of water through artificial recharge with infiltration basins or recharge wells. While basins currently are the most common technique, recharge through wells will increase in the future as good sites for surface infiltration will no longer be available and wells need to be used. The transition from basin recharge to well recharge is already taking place in the Netherlands, where the population density is very high, land is at a premium, and people do not want to change land use and disturb natural ecosystems. Future water management must also emphasize water conservation, which means different things to different people, but can best be defined as minimizing water losses, such as evaporation, transpiration, discharge into oceans or salt lakes, and perched groundwater or other water in the vadose zone, and avoiding such serious deterioration of water quality that treatment becomes too expensive. Treatment and planned use of sewage effluent or other wastewater also will become increasingly necessary. Reuse and recycling are the ultimate forms of resource management, and water is no exception.

Increasing demands for water will lead to fierce competition for water. On a national scale, this can lead to unrest and internal strife. Internationally, wars can erupt. Diplomacy and conflict management will become increasingly important in settling water disputes. Hopefully, countries eventually will have to spend so much money on water projects that there is no money left for war!

FOOD PRODUCTION AND IRRIGATION

The increasing population will also require more food and in many areas this will mean more irrigation. Of course, this presents a collision course because, while more and more irrigation is needed for food production, less and less water will be available for irrigation because of municipal and industrial demands. At the beginning of this century, 90% of all the water use in the world was for irrigation. By 1960, it was ≈80%, currently it is ≈70%, and by the year 2000, it is expected to be ≈60% (Biswas, 1993). Thus, the mandate is clear: more food must be grown with less water. This means more intensive agriculture and use of more fertilizer and pesticides, but this in turn will cause more groundwater pollution with agricultural chemicals (Bouwer, 1990). This is very serious, considering that groundwater is a major water resource and that there is ≈67 times more fresh water stored underground within drillable distance than in all the rivers and lakes of the world (Bouwer, 1978). Most of the fresh water of the world is stored in polar ice caps, but this is not of much use to people. LISA (Low Input Sustainable Agriculture) and concerns about pesticide residues and nitrates in groundwater will be hard to promote in Third World countries, where the main problem often is where the next meal is coming from and not long-term health effects and environmental concerns. In addition to nonpoint source pollution by agriculture, which can involve vast areas, groundwater in the future also will be very much at risk because of tendencies toward groundwater overdraft.

Since irrigation uses so much water, there will be pressure to increase the irrigation efficiency, particularly since the public often perceives irrigation as a wasteful use of water (Bouwer, 1993b). The inefficiency of irrigation is due to deep percolation and tail water runoff losses, which cause individual field efficiencies to be low (<40% with poor design and management). Since these losses are often used again by drainage of groundwater and surface water into streams, by pumping groundwater, or by collecting tail water for irrigation of lower fields, the irrigation efficiencies of large irrigated areas (valleys, districts, or basins) often are much higher than those of individual fields; i.e., >90%. In many large irrigated areas, very little water actually leaves the system, in accordance with the principle that "the upper basin's inefficiency is the lower basin's water source." The main loss in irrigated agriculture is evapotranspiration. This is a consumptive use of water that really cannot be reduced by increasing irrigation efficiency. The only way to reduce the consumptive use of water in irrigated agriculture is to reduce the irrigated area (with possible increases in crop yield per unit water used so that total production remains the same), to grow more cool season crops and fewer warm season crops, to grow more drought tolerant crops, and to grow fewer forage crops and more crops for direct human consumption.

Irrigated areas basically are large evaporation pans where distilled water is returned to the atmosphere in the vapor phase, and salts remain behind in the soil. Thus, salinity must be very carefully managed to ensure a sustainable irrigated agriculture. For coastal areas, salty irrigation return flows can be discharged into the ocean. Where river water is used for irrigation, these salts would have reached the ocean anyway, but diluted with more water. For inland areas, salts can be stored for a while in vadose zones, aquifers, lakes, and evaporation ponds, but eventually salt depositories must be created where salts from irrigated areas can be concentrated and stored forever.

WATER TRANSFERS

As water resources become scarcer, there will be a trend of shifting water use from low economic returns to higher economic returns, which will mean transfers of water from agricultural to urban uses (municipal and industrial). In some countries, water is owned by the state and water transfers can be handled by decree. In other countries, water rights belong to individuals and private interests. Water transfers must then be based on three principles: voluntarism, infrastructure, and third party interests. Firstly, voluntarism is necessary because water cannot just be confiscated from the farmers. There must be an economic incentive. If, for example, farmers can make more money with their irrigation water by selling it to a city instead of using it for growing low value crops, they will be more than willing to sell it. Also, farmers may have both surface water rights and rights to groundwater that they can pump with their own wells. If they can sell their surface water for more money than it costs them to pump groundwater, they can sell their surface water to cities for municipal use and pump groundwater for their own use. Of course, this should be carefully managed to make sure that there is no unacceptable groundwater overdraft. The second principle for successful transfer of water is that the area must have a good infrastructure of water conveyance facilities so that water can be moved around. The third principle is that third party interests must be protected. These interests can include various aspects of the local rural economy, like agricultural jobs and businesses that depend on agriculture. Loss of jobs and business often is offered as a serious argument against transferring water from agricultural areas to urban areas. These economic effects, however, tend to be gradual as water transfers start small and increase slowly. Also, they are not different from the general economic dynamics that are occurring all the time (i.e., layoffs and plant closings by industries large and small, closing of military bases, or urban expansion).

Third party effects also include environmental considerations. For example, if a city begins to divert more water from a stream and returns that water to the stream in the form of sewage effluent, the interests of downstream users of that water must be protected. Affected persons can include farmers that use the water for irrigation, which may require additional N removal from the sewage effluent when the crops are sugar beet (*Beta vulgaris* L.), malting barley (*Hordeum vulgare* L.), alfalfa (*Medicago sativa* L.), or others that cannot have N in the later stages of the growing season. Also, the water should be pathogen free to permit unrestricted irrigation if farmers want to grow lettuce (*Lactuca sativa* L.) and other vegetables consumed raw or brought raw into the kitchen. Recreational uses of the stream also may have to be protected, so that the city must treat the sewage to meet the requirements for primary contact recreation (no pathogens). To protect aquatic life, cities may have to remove more N, organic compounds, metals like Cu, and other chemicals that are toxic to aquatic life. Often, these problems and conflicting interests are resolved by litigation. A better approach is for the cities, farmers, or environmentalists to get together and work out the best scheme of water diversion for urban use, sewage treatment before returning the effluent to the stream, and seasonal storage. In California, water transfers are handled through a water banking system, where farmers can sell their water to the bank and municipalities and other entities in need of more water can buy the water. At the end of the California drought of 1987 to 1993, the water bank handled ≈1000 million m^3 yr^{-1} (Bouwer, 1993b). Of course now that the drought is over, the interest in water banking has decreased.

WATER REUSE

Another water management technique that will become more and more necessary in the future is reuse of wastewater, especially planned and controlled reuse, where the wastewater receives proper treatment to meet the quality requirements for the intended use. Uncontrolled use of sewage effluent for irrigation and other purposes is occurring in many countries, even though it generally means that poorly treated or even raw sewage is used for irrigation of vegetables and other crops consumed raw or brought raw into the kitchen. This is, of course, completely unacceptable from a public health standpoint. If this practice is allowed to continue and spread in the future, there can be major epidemics such as cholera (including the new Bengal strain discovered in Bangladesh in 1992, which is resistant to vaccination), typhoid, hepatitis A, B, and E with the new E-strain having a much higher mortality than A and B, giardia, cryptosporidium, and others, including new ones that are not yet known. Controlled water reuse will be done for various reasons. One is that sewage effluent simply is an important water resource that is needed in water-short areas. With more and more people living in cities, more and more sewage effluent will be produced (megaflows from megacities!) and must be used again to reduce the stress on available water resources. Secondly, reuse may become increasingly attractive to reduce pollution of surface water. For example, in the face of increasingly stringent discharge requirements to protect aquatic life and downstream users of the stream, cities may find it cheaper to treat their sewage effluent for nonpotable use than for discharge into surface water.

While flush toilets and sewer systems are marvels of home and urban sanitation, from a water quality and water reuse standpoint they are great water wasters and polluters. First of all, toilets use a lot of water that is severely degraded by what is flushed down. This water then mixes with the other household wastewater (gray water), which in itself is fairly innocuous and suitable for various reuses, to produce a domestic effluent that is loaded with pathogens that can spread acute infectious diseases and death, with nutrients that can eutrophy surface water, and with toxic chemicals that can have long-term adverse health effects such as cancer. Then, to make things even worse, this effluent is discharged into streams or other surface water where it degrades the quality of much more water that often is the water source for downstream people (dilution is not the solution to pollution!). Even sewage treatment is not the answer because conventional treatment plants do not take out all the contaminants. Thus, end-of-pipe treatment of sewage and disposal into surface water must be more and more replaced by local treatment for reuse and recycling to eventually achieve zero discharge.

Sewage effluent primarily will be used for nonpotable purposes, including industrial uses (power plant cooling, processing plants, or construction), municipal uses (fire fighting and toilet flushing, especially in high rises or industrial buildings), urban irrigation (parks, playgrounds, landscaping, and private yards), agricultural irrigation (including crops consumed raw), environmental applications (wildlife refuges or in-stream benefits), and groundwater recharge. Potable use is also possible, but that usually will be a practice of last resort because of high treatment costs (on the order of \$0.5 to \$1 m^{-3}) and public acceptance, aesthetic, psychological, and cultural or religious reasons (Bouwer, 1993a). If wastewater is used extensively for nonpotable purposes, however, there often will be enough high quality water left for potable uses. The main principle of water reuse is that the sewage effluent must be treated to meet the quality requirements for its intended use. The necessary treatment technologies are available, even to make distilled water out of sewage effluent if the user is willing to pay the price. Because transport of water over long distances can be very expensive, water reuse generally will be concentrated in and around the cities producing the effluent. Irrigation may then require treatment of the effluent to permit unrestricted irrigation, which includes urban irrigation of parks, playgrounds, golf courses, sports fields, and residential yards and agricultural irrigation of fruit and vegetables consumed raw or brought raw into the kitchen. Vegetable growing around cities is very common and economically attractive because of proximity to markets, and sewage effluent should be adequately treated so that farmers can grow such crops.

Use of sewage effluent for cooling water for power plants usually requires treatment to minimize scaling in the pipe system. For a 3810 megawatt nuclear power plant west of Phoenix that is entirely cooled with sewage at a design flow of 3.2 $m^3 s^{-1}$, the effluent first receives conventional primary and secondary (activated sludge) treatment in the sewage treatment plant. It is then transported to the nuclear power plant where it is treated on-site with lime precipitation, trickling filters, and sand filters to minimize scaling in the plant. The cooling water is recycled $\approx$15 times after which it has reached a salt concentration of $\approx$15 000 mg L^{-1}, or half of that of seawater. The brine is then discharged into a 196-ha evaporation lake from which the salts eventually will be removed for disposal as solid waste in designated landfills.

For agricultural irrigation, the effluent first of all has to meet the normal chemical requirements for irrigation water, such as total dissolved solids, Na adsorption ratio, N concentration, Cl^- concentration, and trace element concentrations (Ayers & Westcot, 1985; Bouwer & Idelovitch, 1987). Usually, most sewage effluents from residential areas will meet these standards. If there is a lot of industrial waste going into the sewer system, some chemicals may exceed maximum limits. In that case, source control is necessary to minimize the concentrations of the undesirable chemicals. The most important parameter of sewage effluent for irrigation is its concentration of viruses, bacteria, protozoa, and eggs of parasitic worms (helminths) that can cause diseases in the people consuming the crops or contacting the water (Bouwer, 1993a). For unrestricted irrigation, including urban irrigation and irrigation of vegetables consumed raw or brought raw into the kitchen, the sewage effluent should be treated to remove all pathogens, according to California standards and standards patterned thereafter (U.S. Environmental Protection Agency, 1992). The indicated treatment to achieve this typically is primary and secondary treatment followed by coagulation, sand filtration, and chlorination (Asano et al., 1992) to produce fecal coliform concentrations of essentially zero. This is a high technology and expensive process (about \$200 to \$500 1000 m^{-3}; Richard et al., 1992), however, that may not be suitable for countries with insufficient capital or human resources to build, maintain, or operate such plants. For those countries, the World Health Organization guidelines may be used (World Health Organization, 1989), which allow a maximum fecal coliform concentration of 1000 100 mL^{-1} and up to one helminthic egg per liter. These guidelines can be achieved with low technology treatment systems such as lagooning, if sufficient detention times are used ($\approx$1 mo in warm climates) to get die off or removal of most of the pathogens (Bouwer, 1993a).

Soil-Aquifer Treatment

Another low technology treatment system that can considerably improve the quality of the sewage effluent to the point that it can be used for unrestricted irrigation and most other nonpotable purposes is groundwater recharge with infiltration basins (Bouwer, 1985). The systems then are operated as infiltration-recovery systems to recover the water from the aquifer with wells or drains (Fig. 8–1), and to use the underground formations as a treatment facility. For this reason, they are called soil-aquifer treatment (SAT) or geopurification systems. The water obtained from the recovery systems after SAT usually has a very low suspended solids content, essentially zero BOD, significantly reduced concentrations of N, P, organic compounds, and heavy metals, and essentially zero levels of pathogens (Bouwer, 1985, 1993a) as shown in Table 8–1. Such water can then

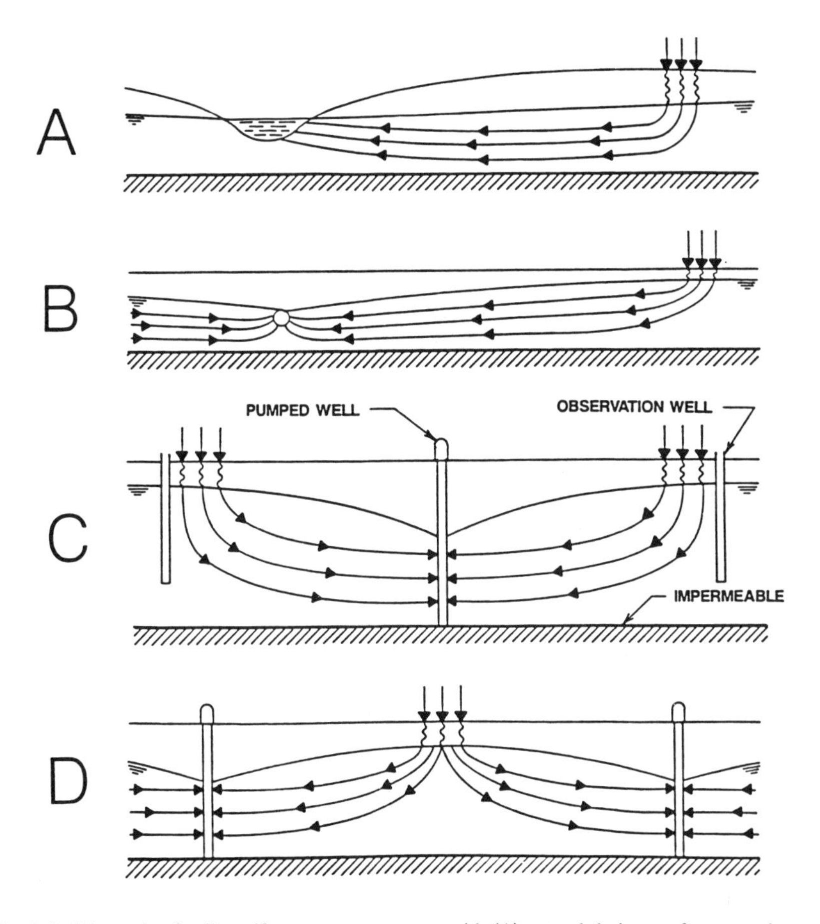

Fig. 8–1. Schematic of soil-aquifer treatment systems with (A) natural drainage of renovated water into stream, lake, or low area; (B) collection of renovated water by subsurface drain; (C) infiltration areas in two parallel rows and lines of wells midway between; (D) and infiltration areas in center surrounded by a circle of wells.

Table 8–1. Quality parameters from Phoenix, AZ, soil-aquifer treatment (SAT) system for mildly chlorinated secondary effluent (activated sludge) as it entered the infiltration basins (left column) and after SAT and pumping it from a well in the center of the infiltration basin area (right column).

	Secondary effluent	Recovery well samples
	mg L^{-1}	
Total dissolved solids	750	790
Suspended solids	11	1
Ammonium NO_3–N	16	0.1
Nitrate NO_3–N	0.5	5.3
Organic NO_3–N	1.5	0.1
Phosphate P	5.5	0.4
Fl	1.2	0.7
Br	0.6	0.6
Biochemical O_2 demand	12	0
Total organic C	12	1.9
Z	0.036	
Co	0.008	
Cd	0.0001	
Pb	0.002	
Fecal coliforms 100 mL^{-1}	3500	0.3
Viruses, PFU/100 L†	2118	0

†PFU, plaque forming units

be used for unrestricted irrigation and most other nonpotable purposes without further treatment. Other advantages of SAT are that the systems are inexpensive and low technology. They are robust and simple to operate, they offer underground storage to absorb seasonal or other differences between supply and demand of water, and they enhance the aesthetics of water reuse by breaking the pipe-to-pipe connection of direct recycling with an underground transport and storage phase.

Some aspects of SAT that need further attention are the health effects, especially if the water after recovery is used for drinking, and the sustainabilty of the system. Most of the underground treatment processes are renewable and should go on indefinitely, but some compounds like phosphate, heavy metals, and some synthetic organic compounds may accumulate in the underground environment. More knowledge is also needed about the role of the clogging layer in infiltration basins and how to manage the basins for minimum infiltration reduction by clogging and maximum quality improvement benefits as the water moves through the clogging layer. More research on control of clogging is also needed for recharge wells, especially dry wells or recharge shafts in the vadose zone, because such wells cannot be pumped to reverse the flow and remove clogging materials.

REFERENCES

Asano, T., D. Richard, R.W. Crites, and G. Tchobanoglous. 1992. Evolution of tertiary treatment requirements in California. Water Environ. Technol. 4(2):36–41.

Ayers, R.S., and D.W. Westcot. 1985. Water quality for agriculture. FAO Irrig. and Drain. Pap. 29. Food and Agriculture Organization, Rome.

Biswas, A.K. 1993. Water for agricultural development: Opportunities and constraints. Int. J. Water Resour. Dev. 9(1):3–12.

Bouwer, H. 1978. Groundwater hydrology, McGraw-Hill, New York.

Bouwer, H. 1985. Renovation of wastewater with rapid-infiltration land treatment systems. p. 249–282. *In* T. Asano (ed.) Artificial recharge of groundwater. Butterworth Publ., Boston.

Bouwer, H. 1990. Agricultural chemicals and groundwater quality. J. Soil Water Conserv. 45(2):184–189.

Bouwer, H. 1993a. From sewage farm to zero discharge. J. Eur. Water Pollut. Control 3(1):9–16.

Bouwer, H. 1993b. Urban and agricultural competition for water, and water reuse. Int. J. Water Resour. Dev. 9(1):13–26.

Bouwer, H., and E. Idelovitch. 1987. Quality requirements for irrigation with sewage effluent. J. Irrig. Drain. Div. ASCE 113(4):516–535.

Postel, S. 1992. Last oasis. Worldwatch Inst., Washington, DC.

Richard, D., T. Asano, and G. Tchobanoglous. 1992. The cost of wastewater reclamation in California. Rep. from Dep. of Civil and Environ. Eng. Univ. of California, Davis.

U.S. Environmental Protection Agency. 1992. Guidelines for water reuse. EPA Manual EPA/625/R-92/004. USEPA, Washington, DC.

World Health Organization. 1989. Health guidelines for the use of wastewater in agriculture and aquaculture. Tech. Bull. Ser. 77. World Health Organization, Geneva.

9 Soil Physics and Groundwater Quality

Jan van Schilfgaarde
USDA-ARS
Beltsville, Maryland

In the first part of *Out of the Earth: Civilization and the Life of the Soil*, Daniel Hillel discusses the difference between the two injunctions to man, expressed in the first chapters of Genesis. In the first, man is appointed to "conquer, subdue, and hold dominion over the earth" and all its living forms. In the second, man is charged with the responsibility "to serve and preserve" the earth, or "to dress it and keep it," depending on which translation of the Hebrew Bible one chooses. I prefer the words "to cultivate and conserve the earth," that is to say, to serve as its steward.

Hillel states that in the time scale of human life, soil is a nonrenewable resource, and so is a mature forest, a river, a lake, or an aquifer. And that is the framework in which I wish to address the topic of this chapter.

Water is indeed the lifeblood of civilization, of all civilizations. As we reflect on the course of our own country, the USA, we cannot but realize that our history was based on a driving force, a credo, of conquering the land and stilling its waters. An example was the passage of the Swamp Act and similar legislation by which Congress induced people to drain, settle and develop marshlands (as in North Carolina and Virginia). I remember the description I read in an old textbook on drainage, of how the early settlers in Illinois and Iowa rode their horses through the mosquito-infested swamps and the horses sank to their bellies in the mud. Now we can drive through the same area and, except for the unusual season of 1993, hardly see any marshes or standing water at all.

Later, irrigation development became dominant in the agricultural expansion of the USA. The Reclamation Act of 1902 was a major stimulus to water development as a basis for settling Western lands. Again, this thrust was impelled by an ethos of development and land ownership that justified the conquest of nature and, especially, its waters.

There seems to have been a parallel development in the philosophy and practice of soil physics. Early on, we had the foundations of our science laid out for us by H. Darcy, E. Buckingham, and F.H. King. They were followed in the 1920s, 1930s, and 1940s by Dr. W. Gardner of Utah and by L.A. Richards. Later still came my own mentor, D. Kirkham, and many others (some still active among us) who made significant contributions to the theory and practice of soil physics. The

Soil and Water Science: Key to Understanding Our Global Environment, SSSA Special Publication 41.

resulting science rationalized the state and flow of water in soils, and set out the principles for measuring soil hydraulic properties and processes. Gradually, the knowledge acquired was put to practical use in the expansion of drainage, irrigation, and groundwater uses.

In recent decades, however, a fundamental change has taken place in our collective ethos. A new philosophy evolved, which we might call the environmental ethic. In part, we ran out of new lands to conquer or new water resources to develop. More importantly, we began to realize that unbridled development is often wasteful, that it may degenerate into misuse. People began to question whether we really know enough to control our scientific and technological thrusts and their ultimate impacts upon environmental and even global biospheric processes. Specifically, what might be the impact of our deeds on water quality in relation to salts, nitrates, and a bewildering array of pesticide residues?

A telling example is the Colorado River water and its salinity. For several generations, the river was untamed and therefore hardly usable. When it was dammed, water quality was taken for granted and the only issue under consideration was water quantity and its supply to potential or actual users. It was not until the late 1960s that the quality issue came into prominence. It quickly became an issue of international contention between the USA and Mexico. Irrigators have long known that the only way to sustain farming under arid conditions is to leach the soil periodically of the accumulating salts and to drain the leachate somewhere. Until recently, the sole concern of irrigators was to rid their land of salts by means of drainage; where the drainage discharged was not their concern. It was someone else's problem.

Belatedly, in the 1970s, we began to realize that the off-site effects of irrigation on water quality can be fully as important as the on-site effects. Concurrently with that issue came competition between agricultural and urban uses of water. The benefit to agriculture of draining the salts to the river entailed extensive damage to downstream users, most notably to urban users (e.g., in Los Angeles) who were forced to replace clogged or corroded water heaters and plumbing fixtures.

Other examples abound. At the Hanford Site in Washington State there is much concern lest radioactive wastes leak into groundwater, and whether or when some of it might end up in the Columbia River. Other sites give us cause for concern, and will continue to do so for a long, long time, not just over radioactive waste, but over many other sorts of toxic residues.

So, increasingly, water quality has become a major issue in the protection of our environment. Consequently, soil physics has shifted its emphasis from an assessment of quantitative water flow to a combined study of transport processes for water and various other mobile constituents. The fate of solutes in such processes as diffusion, dispersion, dissolution, precipitation, volatilization, phase changes, and adsorption became the focus of research and monitoring. The entire field of soil physics thus became much more interesting; that is to say, much more complex, challenging, controversial, and relevant.

We have had our disappointments, too. I read the recent proceedings edited by van Genuchten and colleagues on predicting the hydraulic properties of unsaturated soils, and noted that not a great deal of progress has been made in this difficult area since the seminal contributions of Millington and Quirk, and of Brooks and Corey. We have achieved a qualitative understanding of the transport of pollutants, but still find it difficult to make quantitative predictions of arrival times and concentrations.

Along the way, we have also encountered the problem of spatial variability. We have learned much about that, too, and can now describe that variability in geostatistical terms. Yet, we still cannot make good measurements of the transport properties in the field without expending inordinate time and resources in each case. Added to the problem of spatial variability is the no less daunting problem of bypass flow. Together, these phenomena create a rather disturbing situation when we seek to apply our conceptual knowledge to the solution of practical engineering problems in the field. At the same time, our need to manage larger and more complex domains and ecosystems has become more urgent than ever. For example, we understand in principle that NO_3 fertilizer residues often migrate to groundwater, but have difficulty in deriving a mass balance to quantify the fate of N in the soil. Similarly, we know that pesticides may contaminate groundwater and surface water supplies, yet we cannot provide quantitative predictions.

I am reminded that it is well over a century ago that E.W. Hilgard wrote that if the irrigation farmers in California wish to avoid the problems of salinity encountered in the Old World, they must either change their ways or pray to the good Lord to waive the laws of physics for them (Hilgard, 1886). Our recent experience with salinity teaches us that the problem is even more complicated than Hilgard could have envisioned. It is not simply a matter of total salt content, but also of the toxic concentrations of specific trace constituents such as Se and B. It was just such toxic elements that were implicated in the famous Kesterson Reservoir debacle, described so poignantly in Hillel (1992).

So the question posed to scientists now is: Can we translate our knowledge into practical solutions to the real-world problems faced by contemporary society?

Farming families need and now demand safe drinking water. Hence, it was farmers who asked for the legislation in the State of Iowa that led to the establishment of the Leopold Center for Sustainable Agriculture, rather than the faculty of Iowa State University. More and more, initiatives to protect or restore wildlife come from citizens, including farmers, and that in spite of the fact that legislation to allocate more water for stream and wetland maintenance and for protecting migratory birds may come at the expense of water for agriculture.

A new ethic is aborning: sustainability, or stewardship. The terms are not precisely defined, but the concepts underlying them are important. The objectives of such stewardship are often in conflict with short-term economic goals and, in some cases, with perceived private property rights. Nevertheless, this ethic is real and dynamic, and it seems destined to grow. It is exemplified by the decision to restore the Everglades of southern Florida as much as possible to its original ecology (Hillel, 1992). It is still unclear whether a viable agriculture can be maintained in southern Florida without jeopardizing the area's ecology.

Can soil physicists rise to the new challenges and lead the move toward better stewardship of land and water?

Our science is now more sophisticated than ever before, not the least thanks to the availability of computer power. That is not an unmixed blessing, however.

Back when scientists and engineers carried slide rules, our computing power was strictly limited, but we at least had a feel for the numbers we were calculating. Before we solved a problem then, we needed to estimate the answer so we could then check our estimate against the answer provided by the slide rule. Now, however, we punch a few buttons on a computer and we may come up with a number precise to the 17th place that is complete nonsense, but it came out of the computer so it is deemed to be right. In the transition, therefore, we may have lost our feel for the system we are attempting to manage. So let us use the computer (or computer modeling) as a tool, not as a crutch.

The climate in which we practice our science is changing. Automatic financial support for land grant universities and for agricultural research stations can no longer be taken for granted. Nor is it sufficient to dazzle our colleagues with wonderful solutions to differential equations in order to win recognition and promotion. Now we must demonstrate that we have a product to deliver that can serve our funding clients to obtain the necessary support.

A good example of the need to serve a useful end is what happened in the San Joaquin Valley in California. The U.S. Government spent more than $30 million in 5 yr on a research program to study the drainage problem after Se toxicity came to light. The research program culminated in a report that avoided recommending politically nonpalatable actions. Consequently, the public did not get the answers truly needed. Now we have a similar situation in Florida, and I hope the scientists concerned will have the courage to formulate a more direct set of proposals so that policy makers can act effectively on the basis of scientific facts.

In conclusion, I wish to restate that soil physics has grown into a mature profession and must now make relevant contributions and carry greater responsibility than in the past, to guide society in the increasingly complex and fateful task of soil and water management. We must point the way toward sustainable stewardship of the land that sustains us.

REFERENCES

Hilgard, E.W. 1886. Irrigation and alleali in India. Report to the President, Univ. of California Bull. 86. California State Print. Office, Sacramento.

Hillel, D. 1992. Out of the earth: Civilization and the life of the soil. Univ. of California Press, Berkeley.

10 International Agricultural Research and Development: Future Challenges

N. C. Brady

United Nations Development Programme and
the World Bank
Washington, DC

I am pleased and honored to focus briefly on a subject of vital concern to all humanity, but especially to those living in the less well developed areas of the world. All of us are being reminded daily of how important water is to our well being. But those living in developing countries have even greater concerns. They are poor, hungry, and lack the most basic necessities of life. Their future is clouded by the fact that their population numbers will likely double in the next 35 yr. Their numbers are now increasing at the rate of ≈85 million a year, and they are seeking not only food and sustenance for their growing populations, but hopefully a better life for them. They want to have clean water to drink. They want sanitation conditions that only water can provide. They want their industries to grow and know that this is possible only if water is available. Above all, they must provide food for their rapidly expanding populations, and this takes water, even more than has been used in the past.

THE GREEN REVOLUTION

To help guide us in the future, a brief review of agriculture's progress in the past 30 yr is appropriate. We have witnessed during that period a revolution that in terms of numbers of lives affected truly dwarfs even the industrial revolution. During a period of unprecedented increases in human populations, developing countries as a group actually increased their per capita food production. Much of this increase was due to an expansion of arable land areas. Forests and grasslands were cleared and agricultural crops were grown. But most of the increase came from increased yields per hectare. Science gave farmers new tools to stimulate this revolution, but expansion in irrigation and the use of fertilizers and wise public policies were needed to take advantage of these tools. Since 1950, the world's irrigated area increased by 2.5 times (Petit, 1993). This expansion in irrigation is thought to have accounted for >50% of the increases in global food production (Crosson & Rosenberg, 1990). Furthermore, most of the increases in irrigated land took place in developing countries led by India and China.

Soil and Water Science: Key to Understanding Our Global Environment, SSSA Special Publication 41.

Greater efficiency in the use of water in rain-fed areas also contributed to the success of the green revolution. High yielding cereal varieties along with increased supplies of plant nutrients from chemical fertilizers and organic residues not only stimulated crop production, but increased the efficiency of water usage and reduced water runoff and erosion.

The green revolution also involved some environmentally unfriendly aspects. For example, chemical pesticides that were used to control insect pests and diseases were later found to be deleterious to the well being of humans and other creatures. Water pollution from these chemicals and from unusually high fertilizer applications had some serious environmental consequences. Likewise, excessive salt build up in some irrigation areas was damaging to crop production. The green revolutions has been unfairly blamed for increases in soil erosion. In fact, it has done just the opposite. By intensifying food production in level to rolling areas, it has reduced the need for the steeper slopes of the uplands to produce the needed food.

POTENTIAL FOR THE NEXT 30 YEARS

Unfortunately, the probability of an extension into the future of past successes relating to irrigation seems remote. The best sites for water storage from the standpoint of economics and environmental concerns have already been developed. Future systems will cost more and will receive determined opposition from both the environmentalists and the people native to the areas to be submerged by the water impoundment. The growing opposition of native Americans and environmentalists to expansion of the James Bay hydro-electric plant in Quebec is a case in point (Mitchell, 1993). Enforced displacement of native populations to accommodate new dams is being resisted, as are ecological changes that water impoundment dictates.

Curtailment of expansion of land under irrigation is affected by other issues. There is increasing competition for water use by other development sectors. In developing countries, 80% of the water removed from the surface and underground is used for agriculture. Users often pay little if any for the use of the water. Some 1 billion people in developing countries lack access to potable water, and 1.7 billion lack adequate sanitary conditions (Petit, 1993). This accentuates communicable disease outbreaks. Furthermore, these deficiencies have economic consequences for agriculture as shown by Peru's sharp drop in agricultural exports and tourism during the Cholera epidemic of 1990–1991. Similar constraints pertain to industrial development. Agriculture's monopolistic low cost use of irrigation water in developing countries will probably be increasingly challenged as it is being challenged in this country.

In some areas, excessive pumping for irrigation from underground aquifers has lowered water tables to make such withdrawals economically and environmentally unsound. Likewise, poorly designed irrigation and drainage systems have resulted in excess salt build up that has drastically reduced the crop production potential of these systems. The build up of silt in existing reservoirs will also reduce the expected life to those storage areas and in turn reduce the water being held for irrigation. Each of these three developments will probably reduce the area of land for which given systems can provide irrigation water.

The boost in yields made possible in the past 30 yr by fertilizer applications also will probably not be matched worldwide in the next 30 yr. Fertilizer use rates in parts of Asia already exceed those in North America. While modest increases in fertilizer use can be expected world-wide, only in sub-Saharan Africa where fertilizer use is still very low could one expect marked responses from added plant nutrients.

Another important source of increased food production is the traditional expansion of land under cultivation, merely clear the forests and the grasslands. This provided an important component of food production increases in the past 30 yr. Unfortunately, most potentially productive lands have already been cleared and are being cultivated. In fact, many areas not well suited for cultivation have been cleared. These areas should be allowed to revert to their natural vegetation and no longer be cultivated. The net effect on food production of pressures to bring some lands into agriculture and to release others for forests and natural grasslands will be less pronounced in the coming decades than it has been since 1960.

Increased food production in the years ahead will obviously come from higher yields on existing farm lands. In turn, these yields must be produced using practices and systems that are much more environmentally friendly than are those we have been using. Greater attention must be given to long term sustainable productivity and to the protection and conservation of the natural resources used to produce the food. Water resource management will play a major role in achieving this sustainable objective.

ESSENTIALS FOR SUCCESS

Sustainable food production increases of the future will be determined by two primary developments: (i) new science-based and environmentally sound technologies and systems, and (ii) policies and mechanisms that make it profitable and acceptable for farmers to use these technologies. In other words, we must know what to do and the farmer must benefit from doing it.

There are a myriad of opportunities for science to help us improve water resource management, while simultaneously increasing food production. Only a few will be identified.

Genetic Improvement

Agriculture's interest in soils is to produce plants. In turn, most soil and water management systems involve in some way plants or their residues. Soils don't conserve themselves. Plant cover, root proliferation, crop residues and the cycling of plant nutrients are provided by plants. For that reason, plant improvement and modification must be high on the list of those who want to improve water resource management. Having pest resistant crop varieties that will produce well on excessively acid, saline, or alkaline soils or that will tolerate excesses and deficiencies of water is important for the future, particularly in low income countries. Changing the plant may be less expensive than changing the environment in which it grows. Consequently, the plant breeder, the geneticist, and the biotechnologist must be members of the water resource management team for the future. The introduction of genes and combinations of genes from one species or genera to another may well be among the tools most needed to help the poor countries economically increase their food producing capacities.

Crop and Residue Management

The management of growing crops and their residues largely controls soil and water management relationships. Minimum tillage that leaves the soil covered with crop residues has revolutionized conservation farming in the United States. It has brought about significant reductions in runoff and soil erosion and has increased water infiltration into the soil. Research is needed in developing countries to better understand how similar systems could be put to work there. Recognizing that such research may well have been carried out in overseas research stations, care must be taken to repeat the research on farmers fields. Often farmers reject researchers' systems because they do not meet their needs.

An example of such rejection is seen in the semiarid tropics of India. I have seen research plots that show very clearly the essentiality of crop residues in enhancing water infiltration, in reducing runoff and erosion, and in increasing crop yields. But in farmers' fields outside the researchers' plots one sees little evidence of this remarkable finding. There are no residues, the soil is bare, precipitation runs off, the soil erodes, and yields are low.

An explanation for this anomaly is not the ineffectiveness of the extension or technology transfer system. It is that the farmer cannot meet his or her needs and that of the family when using the so called superior system. The crop residues that are central for this superior system are more badly needed by the farmer and his family to provide fuel to cook their food or to provide feed for their animals than to protect the soil. In the mind of the farmer, his or her system is best.

This situation calls for collaborative research involving biologists, social scientists, and the farm family. Systems that meet the farmers' needs must be created keeping the conservation and sustainability goals in mind. Community projects to supply ample fuel wood may be more effective in convincing the farmers to adopt the soil conserving practices in the semiarid areas of India than all the educational efforts available to them.

Crop management systems that encourage efficient use of water and plant nutrients must also receive attention. For example, multiple cropping systems using crops that differ in their growth patterns, heights, and rooting systems need to be examined to realize better use of water. Such combinations might well be more acceptable for many nonmechanical farms than they are in the USA. Agroforestry systems that use perennial woody species are worthy of experimentation for yield purposes as well as water use efficiency.

Environmental Burden of Pollutants

Developing country farmers share our concerns for polluted water whether it be from domestic or industrial wastes or from upstream farm practices. Research is needed to better quantify the extent to which water borne pollutants are serious problems in the low income countries. Anecdotal evidence suggests that such problems exist. Rules and regulations governing chemical use and disposal are often lacking or unenforced in some countries. The presence in irrigation water of heavy metals, organic wastes, pesticides, and even toxic levels of plant nutrients such as nitrates must be ascertained and monitored. Likewise, the salinity level of irrigation water must be known and if possible, regulated.

Pollutants of agricultural origin must be dealt with by agriculturalists. It is increasingly evident, for example, that these are more environmentally sound methods of controlling insect pests and diseases than the use of chemical pesticides. Integrated pest management systems (IPM) that rely strongly on natural enemies of the pests and on cultural management methods are already in place in some developing countries. The world renowned rice IPM program in Indonesia that is spreading to other Asian countries is a good example. Scientists must help create similar systems for other crops and encourage their use. The creation and use of effective IPM systems must receive high priority during the coming decades.

The build-up of excess salts in irrigated soils results from salt-laden waters coming from upstream systems. Currently >1 million hectares of land each year are being subjected to serious salinization (Maurits la Riviere, 1990) Most of this salinization is due to ineffective or unused soil drainage systems. While at some locations specific research may be needed to identify steps to be taken, in most cases, the solutions are known and simply must be put into practice.

Efficiency of Water Use

The efficiency of use by plants of irrigation water is abysmally low, only ≈37% on a global basis (Maurits la Riviere, 1990). Research has created micro-irrigation techniques that greatly increase this efficiency. Water is delivered directly to plant roots through perforated pipes. Trickle or drip irrigation systems also provide a similar effect with low water losses to evaporation. Such systems are in use in the Middle East as well as in the USA and other industrialized countries. Applied research is needed to evaluate the potential for such water saving practices in developing countries.

Low water use efficiency is also a critical factor in rain-fed agriculture. Soil cover with crop residues reduces evaporation losses leaving soil moisture for later crop absorption. Likewise, plant nutrient levels sufficiently high to encourage optimum plant growth will ensure high levels of crop production per unit of water use. The principles governing this relationship are well known. They merely need to be applied in adaptive research in the developing countries.

Appropriate Technologies for Soil Erosion Control

Other innovative low cost means of reducing excessive runoff and soil erosion must be developed and thoroughly tested. For example, on some sloping lands in the South, the use of the contour tillage and farming systems that have proven to be effective in the North are simply too expensive and machinery-intensive for developing country farmers. Possibilities of using vegetative barriers placed on the contour as a foundation for what is, in effect, contour farming, must be more thoroughly explored. Vetiver grass, which has its champions as well as its opponents, should be tested in areas where competition for moisture is not intensive. Other such living barriers should be sought and their effectiveness evaluated.

The so-called *alley cropping* techniques that use herbaceous species as borders for cross slope strips of cultivated crops is another example of a farming system with some opportunity for reducing soil erosion and runoff. Such systems must be evaluated not only for their biological potential, but for their acceptability by the farmers and their families.

Nutrient Cycling

The efficiency of use of plant nutrients by cultivated crops is notoriously low worldwide. To attain economically optimum yields of foodcrops, fertilizer application rates far in excess of levels taken up by the plants are applied by farmers. The nutrients not used by the crop plants are commonly leached from the soil, volatilized into the atmosphere, or fixed into unavailable forms in the soil. Such practices are unacceptable in the South for both economic and environmental reasons.

Research must be implemented to better cycle nutrients in the crop growing process. Modest rates of chemical fertilizers should be used in combination with organic residues or manures to enhance such cycling. The inclusion of N fixing legumes in the cropping cycle with sufficient P fertilizers to enhance legume growth is a time honored practice that must be more thoroughly explored. Likewise, research on the place of cover crops that can use residual fertilizers must be expanded. The farmers must be partners in the research process to enhance the practical usage of whatever practices are developed.

The role of soil organic matter in nutrient cycling is another critical aspect of soil stability that must be further explored under farming systems of developing countries. The higher soil temperatures and sometimes heavy precipitation in the tropics provide conditions sufficiently different from those found in the North to justify a reexamination of the truisms applicable to temperate areas. It is likely that soil organic matter may play an even more significant role in nutrient cycling in the tropics than that pertaining to the North.

Multifactor Approaches

It is neither feasible nor necessary to carry out applied water resources research under the many different ecological conditions prevalent in developing countries. If every practice and principle found satisfactory in the ecological and social regions of the North must first be tested by researchers in developing countries the progress of development would be stymied. There simply is not enough money or time to implement such research. A viable option to countless field trials is the use of the emerging systems of crop and natural resource models being developed and evaluated around the world.

These models make it possible to obtain reasonable assessments of the effects of practices and combinations of practices on crop performance and even on soil erosion. Effects of climate differences and soil characteristics are taken into consideration by these models. The practice of involving developing country scientists in collaborative research networks designed to perfect and evaluate these models should be continued. Researchers in the North and South can be mutually reinforcing and problems of the South can be approached logically. The low income countries simply do not have the financial resources to implement even the adaptive research that successful modelling research can well replace.

Multifactor systems research by its very nature demands multidisciplinary approaches. Soil and plant scientists must join with entomologists, plant pathologists, engineers, economists, and social scientists, as well as with farmers and their organizations in investigative teams to plan and implement this research. Such research should not be any less rigorous, however, than other research of a disciplinary nature. Efforts should be made to attract the best minds from each discipline to work as interdisciplinary team members. We should resist the temptation to accept watered down research quality in our efforts to be certain that every discipline is represented.

Farmer Incentive Enhancement

Perhaps no area of research is more critical for the success of improved water resource management than that relating to farmers incentives. Unfortunately, too often we ignore the role that the farmer must play in the win-win game of producing more food, while ensuring environmental quality enhancement. Systems are worked out by the biologist, the environmentalist, and the educator and they are all disappointed that the poorly educated, stubborn farmer ignores their best efforts. We must work with farmers to better ascertain what is needed to get them on board or, in contrast, to get us on board to help the farmers do their job while meeting their own basic needs.

Some of this research is related to price policies and some to pure economic incentives. Subsidization of chemical pesticides or of slash and burn clearing of tropical forests for commercial ranching make it difficult to encourage more environmentally sound practices. Research is needed to identify policy and institutional changes that will permit farmers to make more money by adapting sustainable development practices than by not doing so.

More research is needed to identify some desired policy changes. Land tenure systems that give greater security to those who till the land will encourage longer term sustainable practices over those that maximize only current yields. Pricing policies that assure low food prices for consumers at the expense of the food producers gave little incentive for the adoption of input-requiring farming systems. Steps must be taken to remove policy and institutional constraints to sustainable food production practices and systems.

Farmer participation is a must in setting the research agendas relating to farmers' incentives. Development agencies have consistently developed programs and policies that were not successfully implemented, often because the programs and policies did not help the individual farmers.

INSTITUTIONAL COLLABORATION

New collaborative relationships must be used in implementing research that will stimulate future food production in low income countries. First, scientists in the North from both private and public sectors must forge concrete collaborative relations with counterparts in the South. Interaction must be sought among researchers from U.S. universities and private companies with counterpart scientists in international and national centers in the South. The application of sophisticated biotechnology research tools to the problems of third world agriculture must receive high priority. Likewise, the creation and wise use of crop and natural resource models must be a shared responsibility. The use of modern information systems to enhance communications and speed up the research and technology transfer must be a high agenda item.

A second general interinstitutional approach relates to the nongovernmental organizations (NGO). It is increasingly obvious that NGOs may offer more effective linkages between researchers and farmers than the traditional public sector extension systems. Farmer confidence in NGOs is often stronger than with traditional extension programs. Furthermore, the NGOs provide mechanisms to give farmers a better feeling of co-ownership of the development process. If it is theirs, they are more apt to respond positively to proposed courses of action than they have in the past. NGOs and their representatives must be a part of the research and development complex that attacks soil and water management problems.

SUMMARY AND CONCLUSIONS

We are at a crossroads of the agricultural development process. Practices and policies that proved successful in the past may not lead to the sustainable development goals humankind has set for us. We can no longer cut down our forests or burn our grasslands for agricultural use. We cannot depend on the rapid expansions of irrigation and fertilizers that helped fuel the green revolution of the 1970s and the 1980s. We must take steps to reduce the dangers of soil and water pollution and the damage to human health that pesticides and excess nitrates represent.

This leaves us with two primary means of stimulating food production in the decades ahead: (i) generate through science environmentally sound technologies and systems that enhance food production; and (ii) develop and use policies and institutions that assure benefits to farmers and their families. Scientists and educators from the North can best assist this process by developing collaborative linkages with counterparts in the South and with those whom we try to serve, the farmers.

REFERENCES

Crosson, P.R., and N.J. Rosenberg. 1990. Strategies for agriculture. p. 73–83. *In* Managing planet Earth. Readings From Scientific American Magazine. W.H. Freeman & Co., New York.

Maurits la Riviere, J.W. 1990. Threats to the world's waters. p. 37–48. *In* Managing planet Earth. Readings From Scientific American Magazine. W.H. Freeman & Co., New York.

Mitchell, J.G. 1993. James Bay: Where two worlds collide. p. 66–75. *In* Water: The power and turmoil of North American fresh water. Spec. ed. National Geographic, Washington, DC.

Petit, M. 1993. The water resource management policy paper. *In* Report given at Environmental and Sustainable Development Conf. at the World Bank, Washington, DC. 30 Sept.–1 Oct. World Bank, Washington, DC.

APPENDIX

BIOGRAPHY

Dr. Daniel Hillel

A native of Los Angeles, CA, Daniel Hillel was taken at an early age to Palestine. He spent part of his childhood in pioneering settlements in the Jezreel and Jordan Valleys, where he acquired a lifelong interest in agriculture and ecology. After the Second World War, he returned to the USA to complete his high school and college education. In 1951, after earning a master's degree in the earth sciences at Rutgers University, he went to Israel to help in the young state's development. He took part in surveying the country's land and water resources and was a founding member of Sde Boker, the first settlement in the Negev Highlands. While there, he conducted scientific studies of desert ecology and hydrology, and was awarded Israel's first doctorate in soil physics by the Hebrew University of Jerusalem. In late 1956 he was invited to participate in land development projects in Southeast Asia, where he worked for two years and had the opportunity to travel widely and study the ecology of the humid tropics.

After a two-year stint as a postdoctoral research fellow at the University of California, Dr. Hillel was appointed head of Soil Technology and later of the Soil and Water Institute of Israel's Agricultural Research Service, in which capacity he initiated studies of soil and water management and played an important role in developing more efficient methods of irrigation. In 1966, he was invited to join the faculty of the Hebrew University of Jerusalem. He served as professor and head of the Department of Soil and Water Sciences until 1974.

In the course of his subsequent career, Daniel Hillel became increasingly involved in international development projects, working with United Nations agencies as well as the International Atomic Energy Agency, the International Food Policy Research Institute, and the International Development Research Centre. He has served as consultant on land and water management in more than 20 countries in Asia, Africa, and South America and conducted joint research in such countries as Australia, Japan, Holland, Belgium, France, Egypt, Pakistan, India, Nigeria, Iran, and Philippines. Dr. Hillel also cooperated widely with colleagues and institutions in the USA, and served for varying periods as visiting scientist at the U.S. Department of Agriculture and at major U.S. universities. In 1977, he took a position as professor of soil physics and hydrology at the University of Massachusetts, from which he retired in 1993.

Dr. Hillel's research has gradually become more comprehensive and oriented to the larger environment, from studies of soil-water dynamics to soil-plant-water relations and to the formulation and testing of mathematical models of water and solute transport in natural systems and the prevention of groundwater pollution. The published results of his research are widely regarded as definitive and are cited extensively in the scientific and professional literature. He was awarded a patent for his invention of a novel method to conserve soil moisture. He was elected fellow of the American Association for the Advancement of Science, fellow of the American Society of Soil Science, and fellow of the American Society of Agronomy; and he served for several years as National Lecturer for Sigma Xi, the national scientific honor society. Among the honors he has received are the Chancellor's Medal for exemplary service at the University of Massachusetts (1982), a Doctorate of Science honoris causa by Guelph University of Canada (1992), and a John Simon Guggenheim Fellowship and grant (1993).

Dr. Hillel has been active in professional affairs, serving at various times as president of the Israel Society of Soil Science, member of Israel's Commission on the Biosphere and Quality of the Environment, Vice President for Soil Physics of the International Society of Soil Science, Chairman of the Soil Physics Division of the Soil Science Society of America, and editor and reviewer of several scientific journals. He has served as consultant to the California Department of Water Resources, the Environmental Protection Agency, the Department of Energy, Goddard Institute for Space Studies, and the Massachusetts Department of Environmental Quality. He also served as science advisor to the Environment Department of the World Bank and took part in missions to various countries to assess the environmental impacts of development programs, drought relief in Africa, irrigation development in South Asia, and water resources development in the Middle East.

In the course of his career to date, Dr. Daniel Hillel has published well over 200 scientific papers and research reports, embodying his contributions to both fundamental and applied aspects of environmental physics. He has authored and edited 16 books. His textbooks on soil physics are widely used by numerous universities and research institutes throughout the world and have been translated and issued in a dozen languages. In recent years, Dr. Hillel has extended the range of his writing beyond the strictly scientific or professional arena, in an effort to address the general lay readers concerned about the environment. His first effort in this direction was his book *Negev: Land, Water and Life in a Desert Environment* (Praeger, 1982). Lately his articles have appeared in such publications as Natural History Magazine and in the National Geographic Society's Research and Exploration Journal. Dr. Hillel's book entitled *Out of the Earth: Civilization and the Life of the Soil* (The Free Press, 1991; University of California Press, 1992) won the first place award of the American Association of Publishers for publishing excellence in the earth sciences and geography. Dr. Hillel's latest book, *The Rivers of Eden: The Struggle for Water and the Quest for Peace in the Middle East*, was published by Oxford University Press in 1994. He is now coauthoring a new book on the global impacts of climate change on agriculture and on natural ecosystems.

Dr. Robert McC. Adams

Dr. Robert McC. Adams retired in September as the Secretary of the Smithsonian Institution. He received his doctorate from the University of Chicago, where he served as a faculty member (in the Oriental Institute, the

Departments of Anthropology and Near Eastern Civilizations, and the Committee on Public Policy) from 1955 until joining the Smithsonian in 1984. Dr. Adams is a member of the National Academy of Sciences and has served (and continues to serve) on numerous committees of the National Academy and the National Research Council. He has been granted honorary doctorate degrees by eight prominent universities, is a Fellow of the American Academy of Arts and Sciences, and was recently awarded the Great Cross of Vasco Nuñez de Balboa by the Republic of Panama.

Dr. Adams has served as visiting professor and lecturer at leading universities in the USA and Europe. He has conducted field research in Iraq, Mexico, Iran, Syria, and Saudia Arabia. Dr. Adams' main professional interests are the agricultural and urban history of the Near East, comparative economic and social history of pre-modern societies, contemporary policies for the support of research, and more recently in the history of technology. He has authored or edited seven books and well over 100 scholarly articles and reviews.

Dr. Montague Yudelman

Dr. Montague Yudelman is Senior Fellow of the World Wildlife Fund in Washington, D.C. He served as the Director of Agriculture and Rural Development at the World Bank from 1972 to 1984, a period during which the Bank greatly expanded its lending for agricultural development.

Dr. Yudelman, a native of South Africa, earned his Ph.D. in agricultural economics from the University of California at Berkeley. His career has included positions at the Food and Agriculture Organization of the United Nations, the Rockefeller Foundation, and Harvard University. Following his retirement from the World Bank, he was appointed Distinguished Fellow at the World Resources Institute. In his various capacities, Dr. Yudelman has undertaken numerous missions to countries in Africa, Asia, and South America. He has published widely on issues involving the impact of technological change on agricultural development.

Dr. John Letey

Dr. John Letey is Professor of Soil Physics at the University of California in Riverside. He earned his Ph.D. from the University of Illinois in 1959, and has been a member of the faculty at the University of California, Riverside, since 1961. From 1975 to 1980, he served as Chairman of the Department of Soil and Environmental Sciences, and was Director of the University of California's Kearney Foundation from 1980 to 1985. He is currently Associate Director of the University of California's Center for Water and Wildlife Resources. He is a Fellow of both the American Society of Agronomy and the Soil Science Society of America.

Dr. Letey's research has ranged from fundamental aspects of soil-water relations (including pioneering contributions to the study of water-repellent soils) to the pressing problems of irrigation and salinization, in California and elsewhere.

Dr. Rattan Lal

Dr. Rattan Lal is Professor of Soil Physics in the Department of Agronomy at Ohio State University, having joined the faculty there in 1987. He had earned his Ph.D. in soil physics from Ohio State in 1968. After spending a year at the University of Sydney, Australia, he joined the staff of International Institute of Tropical Agriculture (IITA) in Ibadan, Nigeria, where he served for 18 years as Principal Soil Scientist. While at IITA, Dr. Lal carried out definitive research on the sustainable management of tropical soils and on the prevention of soil degradation through depletion and erosion.

Dr. Lal is a Fellow of both the American Society of Agronomy and the Soil Science Society of America. He was the recipient of the International Soil Science Award in 1989, and served as President of the World Association of Soil and Water Conservation from 1988 to 1991.

Dr. Harold E. Dregne

Dr. Harold Dregne is Horn Professor Emeritus of Soil Science at Texas Tech University. His special interests are land development and conservation of natural resources in arid regions. His numerous publications included eight books dealing with arid lands, irrigation, salinity, soils, dryland agriculture, and land degradation. He is a former chairman of the Department of Plant and Soil Sciences and former director of the International Center for Arid and Semiarid Land Studies at Texas Tech. He is a fellow of the Soil Science Society of America and the American Society of Agronomy.

Dr. Dregne has served, and continues to serve, as advisor to and member of national and international panels on land management and sustainable development.

Dr. Cynthia Rosenzweig

Cynthia Rosenzweig is an Associate Research Scientist at the Center for the Study of Global Habitability of Columbia University, and a member of the Climate Group at NASA/Goddard Institute for Space Studies. Her areas of specialization are biological and hydrological land-surface processes for general circulation models and climatic and physiological effects of increased CO_2 on regional, national and international agricultural production. She studied at Stanford and at Rutgers Universities and received her Ph.D. from the Department of Plant and Soil Sciences at the University of Massachusetts under the supervision of Professor Daniel Hillel.

Dr. Rosenzweig coordinated the agricultural studies for the Environmental Protection Agency's Report to Congress, *The Potential Effects of Global Climate Change on the United States*. Recently, she led an interdisciplinary study, sponsored jointly by the U.S. Environmental Protection Agency and the U.S. Agency for International Development, on the implications of climate change for international agriculture, including forecasts of global food production, trade, and vulnerable regions. That study enlisted the participation of collaborating scientists in 25 countries. Dr. Rosenzweig has participated in the Intergovernmental Panel on Climate Change Impacts Assessment as an expert contributor and lead author. She also serves as Adjunct Associate Professor of Environmental Sciences at Barnard College.

Dr. Herman Bouwer

Dr. Herman Bouwer is Chief Engineer at the U.S. Water Conservation Laboratory in Phoenix, AZ. He received his B.S. and M.S. degrees from the State Agricultural University of Wageningen in his native Holland, and a Ph.D. in hydrology and agricultural water management from Cornell University in 1955. After several years at Auburn University in Alabama, Dr. Bouwer joined the staff of the newly established U.S. Water Conservation Laboratory as research hydraulic engineer. From 1972 to 1990, he served as Director of the Laboratory.

Dr. Bouwer's main research interest is in groundwater recharge. He has pioneered in developing methods for the purification, storage, and reuse of sewage effluent, and has served as consultant to various agencies in the USA and abroad in this important area. Dr. Bouwer serves as adjunct professor at Arizona State University and the University of Arizona. He has written some 250 research publications, one textbook (*Groundwater Hydrology*), and 10 book chapters. He received the 1984 Tipton Award from the American Society of Civil Engineers for excellence in water management research, the 1985 Area Scientist of the Year Award from the U.S. Department of Agriculture, the 1988 Hancor Award from the American Society of Agricultural Engineers, and an Honorary Life Membership Award from the National Groundwater Association.

Dr. Jan van Schilfgaarde

Dr. Jan van Schilfgaarde, a native of the Netherlands, earned academic degrees from Iowa State University in engineering and soil physics. After serving for 10 years at North Carolina State University as professor and researcher in soil and water engineering, Dr. van Schilfgaarde joined the Agricultural Research Service (ARS) of the U.S. Department of Agriculture. He began as Water Management Specialist, based in Beltsville, MD, and was later appointed Director of the Soil and Water Conservation Research Division. His next assignment was to be Director of the world-renowned U.S. Salinity Laboratory in Riverside, CA. After several years at Riverside, Dr. van Schilfgaarde took a regional administrative post with ARS at Fort Collins, CO, and finally returned to ARS Headquarters at Beltsville as Associate Deputy Administrator for Natural Resources and Systems.

Throughout his career, Dr. van Schilfgaarde has played a leadership role in agricultural water management. He has edited the important ASA Monograph on Drainage of Agricultural Lands, is a member of the National Academy of

Engineering, and has served on several task forces of the National Academy of Sciences. He has been active in the American Societies of Civil Engineering and Agricultural Engineering, as well as in the Soil and Water Conservation Society and the American Society of Agronomy.

Dr. Nyle C. Brady

Dr. Nyle Brady is Senior International Development Consultant at the World Bank and the United Nations Development Programme. He earned a Ph.D. in soil science at North Carolina State University in 1947, and served as a professor and research administrator at Cornell University from 1947 to 1973. During this period, Dr. Brady also served as Associate Dean of Cornell's College of Agriculture and Life Sciences, and authored the world's most widely used college text in soil science.

From 1973 to 1981, Dr. Brady was the Director General of the International Rice Research Institute (IIRI) in the Philippines. In this capacity, he provided leadership to the team of scientists that developed new rice varieties and technologies to alleviate widespread hunger. During the 1980s, he was Senior Assistant Administrator for Science and Technology at the U.S. Agency for International Development. Dr. Brady has been awarded four honorary doctorate degrees. He is a fellow of both the American Society of Agronomy and the Soil Science Society of America, and was elected President of the Soil Science Society for the period 1963–1964. He has served on panels of the President's Science Advisory Committee and was Chairman of the Agricultural Board of the National Academy of Sciences.

Dr. Glendon W. Gee

Dr. Glendon Gee is a Senior Staff Scientist at Battelle Pacific Northwest Laboratories in Richland, WA. A Certified Professional Soil Scientist, he has been with Battelle since 1977. His interests have been in the measurement and prediction of recharge at arid sites and the analysis of water and solute transport in unsaturated soils. He studied physics at Utah State University and received his Ph.D. in soil physics from Washington State University in 1966.

Dr. Gee is a member and Fellow of the Soil Science Society of America. He is also a member of the American Geophysical Union and the American Society for Testing and Materials. Dr. Gee is the author or co-author of >160 scientific publications in the area of soil physics and waste management and has several patents on soil water sensing devices.

Dr. Ralph S. Baker

Dr. Ralph Baker convened and moderated the symposium from which this volume is derived. He is Principal Soil Physicist and the Technical Director for Soil Science at ENSR Consulting and Engineering's headquarters office in Acton, MA. A Certified Professional Soil Scientist, he has >18 years experience in the evaluation of in situ and on-site treatment of wastes in soil and groundwater. He studied environmental conservation at Cornell University and soil chemistry at the University of Maine, and received his Ph.D. in soil physics from the Department of Plant and Soil Sciences at the University of Massachusetts under the tutelage of Professor Daniel Hillel.

Dr. Baker leads ENSR's National Skill Center for Bioremediation and In Situ Processes. He is currently serving as coordinator and primary author of a comprehensive engineering manual on soil vapor extraction and bioventing prepared for the U.S. Army Corps of Engineers. Dr. Baker is the principal investigator for bioventing and multi-phase flow research studies being carried out at universities under ENSR sponsorship. A member of the Soil Science Society of America, he also serves as visiting lecturer for environmental soil physics at both the Amherst and Lowell campuses of the University of Massachusetts.